how to dunk
a doughnut

len fisher
how to dunk
a doughnut
the science of everyday life

Arcade Publishing • New York

FIRST U.S. EDITION 2003

First published in 2002 in Great Britain by Weidenfeld & Nicolson

Library of Congress Cataloging-in-Publication Data

Fisher, Len.
How to dunk a doughnut : the science of everyday life / Len
Fisher. —1st U.S. ed.
p. cm.
Includes bibliographical references and index.
ISBN 1-55970-680-5
1. Science—popular works. I. Title.

Q162.F53 2003
00—dc21   2003052418

Published in the United States by Arcade Publishing, Inc., New York
Distributed by AOL Time Warner Book Group

Visit our Web site at www.arcadepub.com

10 9 8 7 6 5 4 3 2 1

EB

PRINTED IN THE UNITED STATES OF AMERICA

contents

preface to the american edition

When I began to use everyday activities like dunking to show how scientists think about the world, I hadn't expected my activities to be televised live in America and reported in such prestigious places as the *Wall Street Journal* and the *San Francisco Chronical*, nor to receive letters from American school-children wanting help with their school science projects. These things happened, though, and so I was particularly pleased to be asked to prepare an edition of this book on the science of everyday life for an American audience. I have spent many happy hours working with American scientists, scientists from other countries now living in America, and especially American chefs and food writers. I hope they will forgive me for the stories that I tell about them here.

introduction

Scientists, like hangmen, are socially disadvantaged by their profession. People are naturally curious about their work and their motivation for doing it but are rather afraid to ask about the details. The fear in the case of scientists is that the questioner won't understand the answer, and will end up looking foolish. This fear can be so great that guests at parties, having discovered that I am a scientist, usually turn to my wife and ask her what I do, rather than approach me directly.

This book is for them, and for everyone else who wants to know what scientists are really up to. It uses "the science of the familiar" as a key to open a door to science, to show what it feels like to be a scientist, and to view from an insider's perspective what scientists do, why they do it, and how they go about it. I have used this approach with some success in media publicity exercises designed to show that science can be applied to many everyday activities, including cookie dunking, the best way to use gravy on roast dinners, the making and throwing of indoor boomerangs, and the use of physics to improve your sex life. The widespread public interest in these stories has encouraged me to write this book, in which I give the background to the stories and broaden my repertoire to cover the application of science to doughnut dunking, shopping, household jobs, sports, bathtime, and bedtime — in fact, the major activities of an ordinary day.

Science can add much to everyday activities, but it has also gained much from the study of such activities. Among the things that it has gained are the principle of heat convection, discovered by the Anglo-American Count Rumford after burning his mouth on a hot apple pie; the first measurement of the size of a molecule, performed by Benjamin Franklin

x   introduction

after observing the calming effect of dirty wash water on the waves in a ship's wake; and the first estimate of the range of forces between molecules, derived from consideration of the uptake of liquids by porous materials.

Each chapter is built around a familiar activity, and introduces a major scientific concept that is central to that activity. Interwoven with stories of the science are stories of the scientists, who include many of my contemporaries as well as some famous names from the past. Those from the past cannot stop me from telling stories about them. Most of those from the present have seen what I have written and have kindly refrained from censoring it.

The science of the familiar is one of the most effective ways to introduce science to non-scientists. Michael Faraday, the discoverer of electricity, was among the first in the field nearly a hundred and fifty years ago with his popular lectures on "The Chemical History of a Candle," which were packed by London's fashionable elite. Many others have since followed, including myself.

Not everyone has approved. Some of my colleagues feel that, in reporting experiments on something as commonplace as cookie or doughnut dunking, I am running the risk of trivializing science. Others have taken me to task for bringing science into areas where they feel it has no business to be. One newspaper editor even described me as "the kind of expert who cannot look at a plate of fish and chips without dropping a morsel into a handy test tube and jotting down calculations." The writer was displeased with me for treating gravy absorption by roast dinners as a subject for scientific observation, but he unwittingly hit the nail on the head in describing what science is about. Scientists are in the business of trying to understand the world, and understanding can come just as much from the small and apparently insignificant as it can from contemplation of the grand themes. Many artists, writers, and philosophers have likewise found deep significance in some of the seemingly mundane aspects of life.

Scientists see the world around them in scientific terms,

regardless of time, place, or social propriety. This can lead to some unconventional behavior. The nineteenth-century physicist James Prescott Joule selected a picturesque waterfall as a place for his honeymoon, but his choice was dictated by science rather than romance, and he took a thermometer with him so that he could measure the waterfall's temperature and confirm his theory of heat. When a former colleague of mine was caught in a rainstorm, his rational scientific response was to remove all of his clothes and "hang himself out to dry" over his laboratory radiator, in which position he scared the life out of an unsuspecting cleaning person.

In this book the reader will meet many scientists (most of them clothed) from the past and the present, from different cultural backgrounds, and often with very different scientific and social aspirations. All, though, have shared a vision that Nature's beauty is enhanced by scientific understanding, and that such understanding has its own particular beauty, whether it is concerned with phenomena on the grand scale or with the intimacy of everyday, familiar detail. It is the beauty of that familiar detail that, above all, I wish to share.

I could not have done it without the help of many of my scientific friends and colleagues who have taken the time to discuss issues, to read chapter drafts, and to bring their own expertise to bear in correcting errors and adding enlightenment. Those who have made major contributions include (in alphabetical order) Marc Abrahams, Lindsay Aitkin, Bob Aveyard, Peter Barham, Geoff Barnes, Gary Beauchamp, Tony Blake, Fritz Blank, Stuart Burgess, Arch Corriher, Terry Cosgrove, Neil Furlong, John Gregory, Simon Hanna, Michael Hanson, Robin Heath, Roger Highfield, Philip Jones, Harold McGee, Eileen McLaughlin, Mervyn Miles, Emma Mitchell, David Needham, Jeff Odell, Jeff Palmer, Alan Parker, Ric Pashley, Bob Reid, Harry Rothman, Sean Slade, Burt Slotnick, Elizabeth Thomas, Brian Vincent, and Lawrence West. Other names, equally important, will no doubt come into my mind as soon as this book has gone into print.

I would especially like to thank my agent, Barbara Levy, and my editors, Peter Tallack and Cal Barksdale, for encouraging me in this venture and for showing such belief in my ability to carry it through. Most especially, I would like to thank my wife, Wendy, who has read and reread every chapter draft on behalf of the eventual reader and whose perceptive comments have done so much to remove obscurities and to improve readability.

The book is deliberately designed so that each chapter can be either dipped into or read straight through as a story. There were, furthermore, many fascinating byways, entertaining anecdotes, and small points of interest that did not make it into the chapters, usually because they could not be fitted into the flow of the story without disrupting it. I have put these into notes, some of which are scattered through the chapters, but most of which are accumulated at the back of the book. Here the reader will find advice on the best way to eat hot chili peppers, the rules of the Mudgeeraba Creek Emu Racing and Boomerang Throwing Association, and the reason one American state attempted to sue another for the theft of its rain. These and other tidbits are as much a part of the book as are the main chapters, and I hope that the reader derives as much entertainment from reading about them as I have from discovering and writing about them.

Nunney, Somerset

how to dunk
a doughnut

1
the art and science
of dunking

One of the main problems that scientists have in sharing their picture of the world with a wider audience is the knowledge gap. One doesn't need to be a writer to read and understand a novel, or to know how to paint before being able to appreciate a picture, because both the painting and the novel reflect our common experience. Some knowledge of what science is about, though, is a prerequisite for both understanding and appreciation, because science is largely based on concepts whose detail is unfamiliar to most people.

That detail starts with the behavior of atoms and molecules. The notion that such things exist is pretty familiar these days, although that did not stop one of my companions at a dinner party from gushing, "Oh, you're a scientist! I don't know much about science, but I do know that atoms are made out of molecules!" That remark made me realize just how difficult it can be for people who do not spend their professional lives dealing with matter at the atomic or molecular level to visualize how individual atoms and molecules appear and behave in their miniaturized world.

Some of the first evidence about that behavior came from scientists who were trying to understand the forces that suck liquids into porous materials. One of the most common manifestations of this effect is when coffee is drawn into a dunked doughnut or tea or milk into a dunked cookie, so I was delighted when an English advertising firm asked me to help publicize the science of cookie dunking because it gave me an opportunity to explain some of the behavior of atoms and molecules in the context of a familiar environment, as well as an opportunity to show how scientists operate when they are confronted with a new problem.

I was less delighted when I was awarded the spoof IgNobel Prize for my efforts. Half of these are awarded each year for "science that cannot, or should not, be reproduced." The other half are awarded for projects that "spark public interest in science." The organizers have now changed these confusing descriptions for the simple "First they make you laugh; then they make you think."

It was a pleasure, though, to receive letters from schoolchildren who had been enthused by the publicity surrounding both the prize and the project. One American student sought my help to take the work further in his school science project, in which he studied how doughnuts differ from cookies. He subsequently reported with pride that he had received an "A" for his efforts.

This chapter tells the story of the dunking project and of the underlying science, which is used to tackle problems ranging from the extraction of oil from underground reservoirs to the way that water reaches the leaves in trees.

Doughnuts might have been designed for dunking. A doughnut, like bread, is held together by an elastic net of the protein gluten. The gluten might stretch, and eventually even break, when the doughnut is dunked in hot coffee, but it doesn't swell or dissolve as the liquid is drawn into the network of holes and channels that the gluten supports. This means that the doughnut dunker can take his or her time, pausing only to let the excess liquid drain back into the cup before raising the doughnut to the waiting mouth. The only problem that a doughnut dunker faces is the selection of the doughnut, a matter on which science has some surprising advice to offer, as I will show later in the chapter.

Cookie dunkers face much more of a challenge. If recent market research is to be believed, one cookie dunk in every five ends in disaster, with the dunker fishing around in the bottom of the cup for the soggy remains. The problem for serious cookie dunkers is that hot tea or coffee dissolves the sugar, melts the fat, and swells and softens the starch grains in the cookie. The wetted cookie eventually collapses under its own weight.

Can science do anything to bring the dedicated cookie dunker into parity with the dunker of doughnuts? Could science, which has added that extra edge to the achievements of athlete and astronaut alike, be used to enhance ultimate cookie dunking performance and save that fifth, vital dunk?

These questions were put to me by an advertising company wanting to promote "National Cookie-Dunking Week." As someone who uses the science underlying commonplace objects and activities to make science more publicly accessible, I was happy to give "The Physics of Cookie Dunking" a try. There was, it seemed, a fair chance of producing a light-hearted piece of research that would show how science actually works, as well as producing some media publicity on behalf of both science and the advertisers.

The advertisers clearly thought that there would be keen public interest. They little realized just how keen. The "cookie dunking" story that eventually broke in the British media rapidly spread worldwide, even reaching American breakfast television, where I participated in a learned discussion of the relative problems of doughnut and cookie dunkers. The extent of public interest in understandable science was strikingly revealed when I talked about the physics of cookie dunking on a call-in science show in Sydney, Australia. The switchboard of Triple-J, the rock radio station, received *seven thousand* calls in a quarter of an hour.

The advertisers had their own preconceptions about how science works. They wanted nothing less than a "discovery" that would attract newspaper headlines. Advertisers and journalists aren't the only people who see science in terms of "discoveries." Even some scientists do. Shortly after the Royal Society was founded in 1660, Robert Hooke was appointed as "curator of experiments" and charged with the job of making "three or four considerable experiments" (i.e., discoveries) each week and demonstrating them to the Fellows of the Society. Given this pressure, it is no wonder that Hooke is reported to have been of irritable disposition, with hair hanging in disheveled locks over his haggard countenance. He did in fact make many discoveries, originating much but perfecting little. I had to tell the advertisers in question that Hooke may have been able to do it, but I couldn't. Science doesn't usually work that way.

Scientists don't set out to make discoveries; they set out to uncover stories. The stories are about how things work. Sometimes the story might result in a totally new piece of knowledge, or a new way of viewing the nature of things. But not often.

I thought that, with the help of my friends and colleagues in physics and food science, there would be a good chance of uncovering a story about cookie dunking, but that it was hardly likely to result in a "discovery." To their credit, the advertisers accepted my reasoning, and we set to work.

The first question that we asked was "What does a cookie look like from a physicist's point of view?" It's a typical scientist's question, to be read as "How can we simplify this problem so that we can answer it?" The approach can sometimes be taken to extremes, as with the famous physicist who was asked to calculate the maximum possible speed of a racehorse. His response, according to legend, was that he could do so, but only if he was permitted to assume that the horse was spherical. Most scientists don't go to quite such lengths to reduce complicated problems to solvable form, but we all do it in some way — the world is just too complicated to understand all at once. Critics call us reductionists, but, no matter what they call us, the method works. Francis Crick and James Watson, discoverers of the structure of DNA, didn't find the structure by looking at the complicated living cells whose destiny DNA drives. Instead, they took away all of the proteins and other molecules that make up life and looked at the DNA alone. Biologists in the fifty years following their discovery have gradually put the proteins back to find out how real cells use the DNA structure, but they wouldn't have known what that structure was had it not been for the original reductionist approach.

We decided to be reductionist about cookies, attempting to

understand their response to dunking in simple physical terms and leaving the complications until later. When we examined a cookie under a microscope, it appeared to consist of a tortuous set of interconnected holes, cavities, and channels (so does a doughnut). In the case of a cookie, the channels are there because it consists of dried-up starch granules imperfectly glued together with sugar and fat. To a scientist, the cookie dunking problem is to work out how hot tea or coffee gets into these channels and what happens when it does.

With this picture of dunking in mind, I sat down with some of my colleagues in the Bristol University Physics Department and proceeded to examine the question experimentally. Solemnly, we dipped our cookies into our drinks, timing how long they took to collapse. This was Baconian science, named after Sir Francis Bacon, the Elizabethan courtier who declared that science was simply a matter of collecting a sufficient number of facts to make a pattern.

Baconian science lost us a lot of cookies but did not provide a scientific approach to cookie dunking. Serendipity, the art of making fortunate discoveries, came to the rescue when I decided to try holding a cookie horizontally, with just one side in contact with the surface of the tea. I was amazed to find that this cookie beat the previous record for longevity by almost a factor of four.

Scientists, like sports fans, are much more interested in the exceptional than they are in the average. The times of greatest excitement in science are when someone produces an observation that cannot be explained by the established rules. This is when "normal science" undergoes what Thomas Kuhn called a paradigm shift, and all previous ideas must be recast in the light of the new knowledge. Einstein's demonstration that mass m is actually a form of energy E, the two being linked by the speed of light c in the formula $E = mc^2$, is a classic example of a paradigm shift.

Paradigm shifts often arise from unexpected observations, but these observations need to be verified. The more unexpected the observation, the harsher the testing. In the words of Carl Sagan: "Extraordinary claims require extraordinary proof." No one is going to discard the whole of modern physics just because someone has claimed that Yogic flying is possible, or because a magician has bent spoons on television. If levitation did prove to be a fact, though, or spoons could really be bent without a force being applied, then physics would have to take it on the chin and reconsider.

One long-lived horizontal cookie dunk was hardly likely to require a paradigm shift for its explanation. For that rare event to happen, the new observation must be inexplicable by currently known rules. Even more importantly, the effect observed has to be a real one, and not the result of some unique circumstance.

One thing that convinces scientists that an effect is real is reproducibility — finding the same result when a test is repeated. The long-lived cookie could have been exceptional because it had been harder baked than others we had tried, or for any number of reasons other than the method of dunking. We repeated the experiments with other cookies and other cookie types. The result was always the same — cookies that were dunked by the "horizontal" technique lasted much longer than those that were dunked conventionally. It seemed that the method really was the key.

What was the explanation? One possibility was diffusion, a process whereby each individual molecule in the penetrating liquid meanders from place to place in a random fashion, exploring the channels and cavities in the cookie with no apparent method or pattern to its wanderings. The movement is similar to that of a drunken man walking home from the pub, not knowing in which direction home lies. Each step is a haphazard lurch, which could be forwards, backwards or sideways. The complicated statistics of such movement (called a stochastic process) has been worked out by mathematicians. It shows that his probable distance from the pub depends on the square root of the time. Put simply, if he takes an hour to get a mile away from the pub, it is likely to take him four hours to get two miles away.

If the same mathematics applied to the flow of liquid in the random channels of porous materials such as cookies, then it would take four times as long for a cookie dunked by our fortuitous method to get fully wet as it would for a cookie dunked "normally." The reason for this is that in a normal dunk the liquid only has to get as far as the mid-plane of the cookie for the cookie to be fully wetted, since the liquid is coming from both sides. If the cookie is laid flat at the top of the cup, the liquid has to travel twice as far (i.e., from one side of the cookie to the other) before the cookie is fully wetted, which would take four times as long according to the mathematics of diffusion (Figure 1.1).

Figure 1.1: How to Dunk a Cookie.
Left-hand diagram: Disaster — a cookie dunked in the "conventional" manner, with liquid entering from both sides. Right-hand diagram: Triumph — a cookie dunked in the "scientific" manner. The liquid takes four times as long to penetrate the width of the cookie, and the cookie will remain intact so long as the upper surface stays dry.

The American scientist E. W. Washburn found a similar factor of four when he studied the dunking of blotting paper — a mat of cellulose fibers that is also full of random channels. Washburn's experiments, performed some eighty years ago, were simplicity itself. He marked off a piece of blotting paper with lines at equal intervals, then dipped it vertically into ink (easier to see than water) with the lines above and parallel to the liquid surface, and with one line exactly at the surface. He then timed how long it took the ink to reach successive lines. He found that it took four times as long to reach the second line as it did to reach the first, and nine times as long to reach the third line.

I attempted to repeat Washburn's experiments with a range of different cookies provided by my commercial sponsor. I dunked the cookies, each marked with a pencil in five-millimeter steps, vertically into hot tea, and timed the rise of the liquid with a stopwatch. The cookies turned out to be very similar to blotting paper when it came to taking up liquid. Just how similar became obvious when I drew out the results in a graph. If the distance penetrated follows the law of diffusion, then a graph of the square of the distance traveled versus time should be a straight line. If it took five seconds for the liquid to rise four millimeters, it should take twenty seconds for the liquid to rise eight millimeters. And so it proved, for up to thirty seconds, after which the sodden part of the cookie dropped off into the tea (Figure 1.2).

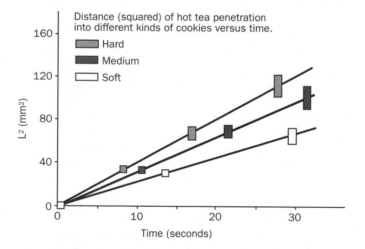

Figure 1.2: Distance (Squared) of Hot Tea Penetration into Different Kinds of Cookies versus Time.
The boxes represent individual measurements, with the lengths of the vertical and horizontal sides representing the probable error in the measurement of (distance)^2 or time respectively.

These results look very convincing. Numerical agreement with prediction is one of the things that impresses scientists most. Einstein's General Theory of Relativity, for example, predicted that the sun's gravitation would bend light rays from a distant star by 1.75 seconds of arc (about five ten-thousandths of a degree) as they passed close by. Astronomers have now found that Einstein's prediction was correct to within one percent. If astrology could provide such accurate forecasts, even physicists might believe it.

That's not the end of the story. In fact, it is hardly the beginning. Even though the experimental results followed the pattern of behavior predicted by a diffusion model, closer reasoning suggested that diffusion was an unlikely explanation. Diffusion applies to situations where an object (whether it is a drunken man or a molecule in a liquid) has an equal chance of moving in any direction, which seems unlikely for liquid penetrating a cookie, since the retreat is blocked by the oncoming liquid. Diffusion models, though, are not the only ones to predict experimentally observed patterns of behavior. Washburn provided a different explanation, based on the forces that porous materials exert on liquids to draw them in.

The imbibition process is called capillary rise, and was known to the ancient Egyptians, who used the phenomenon to fill their reed pens with ink made from charcoal, water, and gum arabic. The question of how capillary rise is driven, though, was first considered only two hundred years ago when two scientists, an Englishman and a Frenchman, independently asked the question: "What is doing the pulling?" The Englishman, Thomas Young, was the youngest of ten children in a Quaker family. By the age of fourteen, he had taught himself seven languages, including Hebrew, Persian, and Arabic. He became a practicing physician and made important contributions to our understanding of how the heart and the eyes work, showing that there must be three kinds of receptor at the back of the eye (we now call them cones) to permit color vision. Going one better, he produced the theory that light itself is wave-like in character. In his spare time he laid the groundwork for modern life insurance and came close to interpreting the hieroglyphs on the Rosetta stone. The Frenchman, the Marquis de Laplace, also came from rural origins (his father was a farmer in Normandy) and his talents, too, showed themselves early on. He eventually became known as "The Newton of France" on account of his incredible ten-volume work called *Mécanique céleste*. In this work he showed that the movements of the planets were stable against perturbation. In other words, a change in the orbit of one planet, such as might be caused by a meteor collision, would only cause minor adjustments to the orbits of the others, rather than throw them catastrophically out of synchrony.

Young and Laplace independently worked out the theory of capillary rise — in Laplace's case, as an unlikely appendix to his work on the movements of the planets. Both had observed that when water is drawn into a narrow glass tube by capillary action, the surface of the water is curved. The curved liquid surface is called the *meniscus,* and if the glass is perfectly clean the meniscus will appear to just graze the glass surface (Figure 1.3).

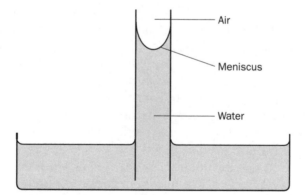

Figure 1.3: Water Rise in a Narrow Glass Tube.

Laplace's (and Young's) brilliantly simple thought was that it appeared as though the column of water was being lifted at the edge by the meniscus. But what was doing the lifting? It could only be the glass wall, with the molecules of the glass pulling on the nearby water molecules. But how could such a horizontal attraction provide a vertical lift? Laplace concluded that each water molecule in the surface is attracted primarily to its nearest neighbors, so that the whole surface is like a rope hammock, where each knot is a water molecule and the lengths of rope in between represent the forces holding the molecules together (Figure 1.4).

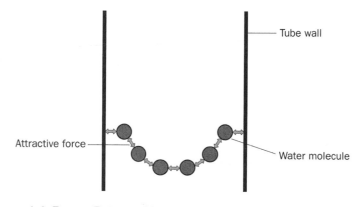

Figure 1.4: Forces Between Molecules in a Meniscus.
The molecules (circles) are held together by attractive forces (arrows). Molecules near the tube walls also experience attractive forces between themselves and the walls.

A hammock supported at each end sags in the middle. A simplistic picture of capillary rise is that the water column is being lifted in a similar manner. More accurately, the forces of local molecular attraction tend to shrink the liquid surface to the minimum possible area. If the surface is curved, the tendency of the surface to shrink (known as *surface tension*) produces a pressure difference between the two sides, just as the stretched rubber surface of a balloon creates a high internal pressure. It is the pressure difference across a meniscus that drives capillary rise.

Laplace was able to use his picture of local molecular attraction to write down an equation describing the shape of a meniscus so accurately that the equation has never needed to be modified since. By thinking about the commonplace phenomenon of capillary rise, he had also unexpectedly found an answer to one of the big questions in science at that time: "How far do the forces between atoms or molecules extend?" Are they long range, like the force of a magnet on a needle, or the force of gravity between the Sun and the Earth? Or are they very short range, so that only nearby atoms are affected? Laplace showed that only very short-range forces could explain the shape of a meniscus and the existence of surface tension. Knowing how many molecules are packed together in a given volume of liquid, he was even able to make a creditable estimate of the actual range of the intermolecular forces. His experience shows that the science of the familiar is more than a way of making science accessible or illustrating scientific principles. Many of the principles themselves have arisen from efforts to understand everyday things like the fall of an apple, the shape and color of a soap bubble, or the uptake of liquid by a porous material. Scientists exploring such apparently mundane questions have uncovered some of Nature's deepest laws.

Laplace and Young showed that the relationship between surface curvature, surface tension, and the pressure across a meniscus was an extraordinarily simple one — the pressure difference across the meniscus at any point is just twice the surface tension divided by the mean radius of curvature at that point. This relationship, which now bears their joint names, shows (for example) that capillary action alone can raise a column of water no more than fourteen millimeters in a tube with a radius of one millimeter. As the tube radius becomes smaller, the water can rise higher in proportion. For a tube one thousand times narrower, the water can rise one thousand times higher.

Such tiny channels are present in the leaves of trees. Nature provides a spectacular example in the Giant Sequoia, found in the Sierra Nevada range in California. The leafy crown of the largest known specimen, the "General Sherman," towers eighty-three meters above the tourists passing below. The water supply for the leaves is drawn up from the soil by capillary action. The menisci of these huge columns of water reside in the leaves, and a quick calculation shows that the capillary channels containing the menisci can be no more than 0.2 micrometers wide — about one two-hundred-and-fiftieth of the diameter of a human hair. The pressure across such a tiny meniscus can support a continuous column of water, of which there are many in the bundles of tubes called the xylem, which run up the trunk below the bark. If the column of liquid breaks, however, an airlock develops at a point where the tube is much wider, and where the new meniscus cannot support anything like such a tall column of liquid. Such breaks are frequent events — the occurrence of each new break is signified by a "click" that can be heard with a stethoscope. Once a column has broken, it seemingly cannot re-form. Eventually, according to the accepted theory, all columns should break and the tree should die. Yet massive trees continue to grow, sometimes for thousands of years, providing botanists and biophysicists with a problem that is a long way from being solved.

The Young-Laplace equation nevertheless provides the only reasonable explanation for the uptake of water by trees. It applies equally to the uptake of coffee by doughnuts, since the coffee is held in place in the porous matrix by the pressure across the meniscus in the smallest of the pores at the upper level of the coffee, just as water is held up in the xylem of a tree by the pressure across the meniscus in the smallest of the pores in the leaves. This leads to the paradoxical conclusion that more finely textured doughnuts should be able to retain more coffee than their coarser-textured cousins, provided that both have the same total pore volume.

The Young-Laplace equation has been applied to many serious practical questions, such as the prevention of rising damp in buildings and the extraction of oil from porous rocks, as well as to the slightly less serious questions of cookie and doughnut dunking. It tells us how far liquids will rise up a tube or penetrate into a porous material, but it doesn't say how fast. This is a key piece of information when it comes to cookie dunking. It was provided by a French physician, Jean-Louis-Marie Poiseuille, who practiced in Paris in the 1830s. Poiseuille was interested in the relationship between the rate of flow of blood and the pressure in veins and arteries. He was the first to measure blood pressure using a mercury manometer, a technique still used by doctors today. He tested how fast blood and other liquids could flow through tubes of different diameters under the pressures that he had measured in living patients, and found that the rate of flow depended not only on the pressure, but also on the diameter of the tube and the viscosity of the liquid (i.e., its resistance to flow: honey, for example, is much more viscous than water). Poiseuille's contribution to science was to describe the dependence of rate of flow on tube diameter and liquid viscosity by means of a very simple equation (the details of which are in the notes to this chapter).

Poiseuille's equation can be combined with the Young-Laplace equation to predict rates of capillary rise. Washburn was the first to do this, producing an equation that predicts how far a liquid drawn into a cylindrical tube by capillary action will travel in a given time. The actual equation is:

$$L^2 = \frac{\gamma \times R \times t}{2\eta}$$

where L is the distance that the liquid travels in time t, R is the radius of the tube, and γ (surface tension) and η (viscosity) are numbers that depend on the nature of the liquid. Washburn's very simple equation predicts an equally simple effect — that to double the distance of travel will take four times as long, and to triple it will take nine times longer; exactly the experimental result that Washburn obtained for blotting paper, and that I obtained for cookies.

In the absence of gravitational effects (which were negligible for both Washburn's and my own experiments), the Washburn equation is extremely accurate, as I found when studying it as a part of my Ph.D thesis some twenty years ago. By timing the flow of liquid down glass tubes (some of them twenty times narrower than a human hair) I found that the equation is correct for tube diameters as small as three micrometers. Such tubes, though, are a far cry from the interiors of blotting paper or cookies. There seemed to be no theoretical reason why an equation derived for a very simple situation should apply to such a complicated mess. There still isn't. No one, to my knowledge, understands why a liquid drawn by surface tension into a tortuous set of interconnected channels should follow the same simple dynamics as a liquid drawn into a single cylindrical tube. All that we can say is that many porous materials behave in this way. The "drunkard's walk" diffusion equation, which predicts a similar relationship between distance penetrated and time taken, may have a role to play. Despite extensive computer modeling studies, though, we still don't have a full and satisfactory answer.

What we do know is that the Washburn equation works. It's not the only equation that works when it's not supposed to. The equation that describes how a thin stream of water dripping from a tap breaks up into droplets, for example, has been applied very successfully to describe the breakup of an atomic nucleus during radioactive disintegration. That doesn't mean that an atomic nucleus is like a water droplet in all other respects, any more than a cookie or piece of blotting paper is exactly like a narrow tube. It simply happens that an equation derived for an idealized situation also applies in practice to more complicated situations, and hence can be used to give guidance and predictions in these circumstances. Such equations are called semi-empirical, and often arise when scientists are in the throes of trying to understand a complex phenomenon. They are most useful at an intermediate stage in the understanding of a problem. When a more complete explanation eventually becomes available, semi-empirical equations are usually discarded, although they sometimes retain a value as teaching instruments.

The Washburn equation, applied to cylindrical tubes, has a sound theoretical basis. Applied to cookies or blotting paper, though, it is semi-empirical. To use it in these circumstances, we need to be able to interpret R. R is a radius, but of what? The best that we can do is to interpret it as an "effective" radius, a sort of average radius of all the pores and channels. One can try to assess the value of this effective radius by measuring as many channels as possible under a microscope and taking an average, but there is a simpler way, using the experimental graph for cookie dunking. The slope of this graph can be used to calculate the effective radius via the Washburn equation. When I tried the calculation, though, the results didn't seem to make sense.

The effective radii of the channels in dunked cookies, calculated from the Washburn equation, were 68, 88, and 110 nanometers for the soft and crumbly, medium, and hard cookies respectively. These radii are very small. The calculated diameters are hundreds, or even thousands of times smaller than the size of the holes that can be seen in a dry cookie under the microscope, which are measurable in micrometers (a micrometer is one thousand times bigger than a nanometer). So what's going on? The answer seems to be that the structure of a wet cookie is very different from that of a dry cookie. In a dry cookie, the starch is in the form of shrunken, dried-up granules. These are quite tiny. In rice (which is almost pure starch), for example, there are thousands of tiny granules in every single visible grain. When these granules come into contact with hot water, they swell dramatically, taking in water as avidly as an athlete during a marathon. As it happens, my colleagues and I had studied the swelling process, which is very important in the preservation, processing, and reconstitution of starchy foods. We held single potato starch granules in water while we gradually raised the temperature of the water, watching what happened through a microscope. At around 60°C the granules suddenly increased their volume by up to seventy times, producing what I subsequently described in a radio interview as the world's smallest potato pancakes.

The starch granules in cookies swell similarly when the cookies are dipped into hot tea. The swollen, crinkled granules become very soft, which is one of the reasons why a dunked cookie puffs up and eventually disintegrates (the other reason is that the fat and sugar "glue" between the granules melts and dissolves). The granules that we were studying became so soft that they could be sucked into glass tubes whose diameters were three times smaller. This deformability seems to be the explanation for the extraordinarily low values of the effective channel radius calculated from the Washburn equation for dunked cookies — the softened granules squeeze up against each other like rock fans at a concert, leaving only the narrowest of gaps in between. In practice, it's just as well. If the pores stayed at their original "dry cookie" size, the Washburn equation predicts that a cookie would fill up with tea or coffee in a fraction of a second, and cookie dunking, unlike doughnut dunking, would become a matter of split-second timing.

As it is, the Washburn equation not only explains why cookies dunked by the "flat-on" scientific method can be dunked for four times as long as with the conventional method — it can also be used to predict how long a cookie may safely be dunked by those who prefer a more conventional approach. Only one assumption is needed — that the cookie will not fall apart so long as a thin layer remains dry and sufficiently strong to support the weight of the wet part. But how thin can this layer be? There was only one way to find out, and that was by measuring the breaking strength of dry cookies that had been thinned down. I consequently ground down a range of cookies on the Physics Department's belt sander, a process that covered me with cookie dust and caused much amusement among workshop staff, who were more used to manufacturing precision parts for astronomical instruments.

Whole dry cookies, I found, could support up to two kilograms of weight when clamped horizontally at one end with the weight placed on the other end. The thinned-down dry cookies were strong in proportion to their weight, and could be reduced to two percent of their original thickness and still be strong enough to support the weight of an otherwise saturated cookie (between ten and twenty grams, depending on the cookie type). All that was needed now was to calculate how long the cookies could be dunked while still leaving a thin dry layer, either in the mid-plane of the cookie for a conventional dunk or on the upper surface of the cookie for a "scientific" dunk. The calculation was easily done using the Washburn equation plus the values of the effective channel radius for different cookies. For most cookies, the answer comes out at between 3.5 seconds and 5 seconds for a conventional dunk, and between 14 and 20 seconds for a "scientific" dunk.

Was there anything else to consider? The only thing remaining was to examine the breaking process itself. The physics of how materials (including cookies) break is quite complicated. The underlying concept, though, is relatively simple (as are quite a few scientific concepts — the expertise comes in working out their consequences in detail). The concept here is that when a crack starts all of the stress is concentrated at the sharp tip of the crack, in the same way that when someone wearing stiletto heels steps on your toe, all of the painful pressure is concentrated at the tip of the heel. If the stress is sufficient to start a crack, it is sufficient to finish the job. That is why brittle materials (including dry cookies) break completely once a break has started. The stress that is needed to drive a crack depends on the sharpness of the crack tip. The sharper the tip, the less stress is needed, in the same way that a light person wearing a stiletto heel can produce as much pain as a heavier person wearing a wider heel. It would seem, then, that even the tiniest scratch could potentially grow into a catastrophic break, no matter how strong the material, so long as the tip of the scratch was sufficiently sharp.

Engineers up to the end of the Second World War knew from practical experience that there must be something wrong with this theory, or else a saboteur could have caused London's Tower Bridge to collapse into the Thames by scratching it with a pin. Even though experience showed that this wouldn't happen, engineers still massively over-designed structures like bridges and ships — just in case. Even so, there were occasions when the theory took over. One such example was when an additional passenger elevator was fitted to the White Star liner *Majestic* in 1928. Stresses concentrated at the sharp corner of the new, square hole in the deck where the elevator was situated drove a crack across the deck and down the side of the ship, where it fortuitously struck a porthole (providing a rather more rounded tip), without which the ship carrying 3,000 passengers would have been lost somewhere between New York and Southampton. In other cases, such as that of the USS *Schenectady*, ships have actually been torn in half (Figure 1.5).

Figure 1.5: Crack Formation on a Grand Scale: the *Schenectady* Disaster.
In January 1943 the one-day-old T2 tanker SS *Schenectady* had just returned to harbor after sea trials when there was a huge bang, and the vessel fractured from top to bottom, jackknifing so that the bow and stern settled to the river bottom while the center rose clear of the water. Photograph reproduced with permission from B. B. Rath, Naval Research Institute.

Sharp corners are now rounded where possible to avoid stress concentration effects. We also understand more about the mechanisms that stop small cracks from growing, which involve plastic (i.e. plasticine-like) flow of the material at the tip of the crack, so that the tip becomes slightly rounded and less sharp. The process can be encouraged by the incorporation of crack-stoppers. These are soft components in a mixed (composite) material whose function is to stop cracks from growing. When a traveling crack hits a particle of crack-stopper, the crack-stopper "gives," turning the crack tip from sharp to blunt and reducing the stress concentration to below a safe limit. The ultimate crack-stopper is an actual hole, such as the porthole in the *Majestic.*

Modern composite materials, such as those used for the manufacture of jet engines, routinely contain "crack-stoppers." Cookies are also composite materials, and also contain crack-stoppers. The crack-stoppers are natural materials like sugar, starch, and (especially) fat, which, although hard, still have some "give." As a result, most cookies are remarkably robust, until they are thinned too far. Then the "graininess" of the cookie takes over. When a cookie becomes as thin as the diameter of the individual grains, the separation of any two grains is sufficient to reveal the void below, and the cookie falls apart.

There is a solution even to this problem — a two-dimensional crack-stopper. That crack-stopper is chocolate, a material that "gives" slightly when an attempt is made to break it, and which can be (and is) often used to cover part or all of a cookie surface.

Our eventual recommendation to the advertisers was that basic physics provides the ultimate answer to the perfect cookie dunk. That answer is to use a cookie coated on one side with chocolate, keep the chocolate side uppermost as you dunk the physicists' way, and time the dunk so that the thin layer of cookie under the chocolate stays dry.

To my considerable surprise, the story was taken up avidly by the media, with the Washburn equation as the centerpiece. The idea of applying an equation to something as homely as cookie dunking made a great hit with journalists. Those who published the equation took great care to get it right; some even telephoned several times to double-check. Only one failed to check, and they got it wrong, provoking the following letter:

Dear Sir,

I think there is something wrong with your cookie-dunking equation. Please send me some cookies for noticing this.

Chao Quan (aged 12)

Unfortunately, by the time the letter arrived, my colleagues and I had eaten all the cookies.

Why should an eighty-year-old equation become the center of a news story? At the invitation of the journal *Nature*, I tried to find an answer. My conclusion was:

Such journalistic excitement over an equation contradicts the normal publisher's advice to authors — that every additional equation halves the sales of a popular science book. Why was this so? Let me suggest an answer, relevant to the sharing of more serious science. Scientists are seen by many as the inheritors of the ancient power of the keys, the owners and controllers of seemingly forbidden knowledge. Equations are one key to that knowledge. The excitement of journalists in gaining control of a key was surely a major factor in their sympathetic promotion of the story. By making the Washburn equation accessible, I was able to ensure that journalists unfamiliar with science could use the key to unlock Pandora's box.

The science of dunking may seem trivial, and in one sense it is. Scientists' questions often seem like a child's idle curiosity, the sort of thing that we should have outgrown when we reached adulthood, so that we could concentrate on more serious things like making money or waging war. To myself and other scientists, though, asking "why?" is one of the most serious things that we can do. Sometimes we try to justify it by practical outcomes. To me, that is a big mistake, whether the outcome is landing a man on the moon or finding a better way to dunk a cookie. The real reason that a scientist asks "why?" is because he or she shares with the rest of the community the most basic of human aspirations — wanting to understand the world and how it works. As members of a thinking species, we all have such aspirations, and express them every time we ponder religious beliefs, or our relationships with other people, or feelings of any kind. Scientists find a similar sense of excitement in addressing a particular area of life's great canvas — the behavior of the material world.

In compensation for the narrowness of our compass, scientists have gotten further in understanding the material world than psychologists, philosophers, and theologians have in their attempts to understand people and their relationships with each other or with the world. It has often happened (as in the case of Laplace) that questions about commonplace phenomena have produced answers to other, and sometimes more important, questions. In the rest of this book, the science of the commonplace is used to open doors for non-scientists. It has often opened doors for scientists as well.

2
how does a scientist boil an egg?

The single egg, in the dark blue egg cup with a gold ring around the top, was boiled for three and a third minutes. It was a very fresh, speckled brown egg from French Marans hens owned by some friends of May in the country. Bond disliked white eggs and, faddish as he was in many small things, it amused him to maintain that there was such a thing as the perfect boiled egg.

Ian Fleming, *From Russia with Love*

Energy, we now believe, is the ultimate stuff of the universe. It comes in various forms — heat, light, microwaves, electricity, and so on. All of these forms have one thing in common: they can be used to move things.

This even applies to domestic cooking, where the energy (usually in the form of heat) that we put into the food does its job by moving the molecules in the food, which wiggle and rearrange themselves to make food more digestible and with a more palatable texture. How it does this was the subject of a very interesting meeting that I attended in Sicily, where scientists and chefs worked jointly to establish the best ways of delivering heat energy to the parts of the food where it mattered. This chapter gives the story of that meeting, and the story of energy itself. For those readers with an eye towards practical value, it also gives the scientific rules for the best way to boil an egg.

James Bond is not the only gourmet to have pursued the perfect boiled egg. If he had driven his 1930 4½-liter gray supercharged Bentley coupé up the tightly folded mountain road that scars the eastern flank of Sicily's Monte San Giuliano, the fluttering growl of its twin exhausts would eventually have echoed from the ancient walls of the village of Erice, rumored to be the former headquarters of the Mafia. These days, Bond would have encountered a different sort of mafia — the gastronomic kind, an international group of chefs and scientists who meet every two years in the Ettore Majorana Center for Scientific Culture to look for ways of using science to extend the horizons of gastronomy. There, in the year 1997, Bond would have found the answer to his quest.

Bond would have been seventy-nine years old; ten years younger than Nicholas Kurti, the former Oxford physics professor who inspired the meetings. Nicholas, at eighty-nine, was still looking for new challenges with an energy that was a tribute to his lifetime's devotion to the pleasures of the table. I had traveled with him from England, puffing in the wake of his small, balding figure as we raced across the concourse of Milan airport to catch a connecting flight to Palermo. His progress seemed unimpeded by a backpack full of thermocouples and recorders for following the temperature changes in food as it was cooked. Nicholas was fond of declaiming that "we know more about the temperature distribution in the atmosphere of Venus than we do about the interior of a soufflé," and this meeting was an opportunity to correct the balance.[1]

It was not Nicholas's first venture into combining science with cooking. He was one of the first television cooks in the U.K., presenting as early as 1969 on black-and-white television a live program called *The Physicist in the Kitchen*, in which he produced some surprising variants on traditional cooking methods. He used a hypodermic syringe, for example, to inject brandy directly into hot mince pies so as to avoid disturbing the crust. He also demonstrated an original technique for making meringues, where he put dollops of creamy meringue mixture onto plates in a vacuum jar and then turned on the pump. The dollops foamed up to produce meringues that were as hard and brittle as any prepared in an oven, but which took a quarter of the time to make and melted in the mouth.

Nicholas was a low-temperature physicist, famous for once

---

[1] Nicholas Kurti, CBE, FRS, died in November 1998, shortly after his ninetieth birthday. This chapter is dedicated to his memory.

having held the world record for the lowest temperature ever achieved in the laboratory. His fame among scientists proved useful when it came to promoting a series of meetings on the science of gastronomy, conceived as a result of a conversation between the San Francisco cooking teacher Elizabeth Thomas and an Italian scientist who happened to be attending a meeting with Elizabeth's husband at the Ettore Majorana Center, a set of converted monasteries in Erice, on quite a different subject. The director was keen on the idea, and promptly asked Elizabeth to organize such a meeting. Elizabeth suggested that Nicholas, an old friend and a leading figure at many Erice meetings, would be the ideal scientific link. The only problem was the title. A series on "The Science of Cooking" seemed rather out of keeping in a place used for hosting discussions on major questions like "Planetary Emergencies" and "The Mathematics of Democracy." Nicholas, ever the pragmatist, suggested the more impressive title "Molecular and Physical Gastronomy," and the Erice series of meetings was born.

The Majorana Center turned out to be an excellent environment. The meeting room, which holds about forty (the upper limit for our meetings), lies on one side of a flagstone courtyard. On the opposite side of the courtyard is the old monastery kitchen, now modernized so that ideas arising in the course of the meeting or those proposed beforehand could be tried out. The meeting title has also turned out to be a good one, and has been widely adopted outside the confines of Erice. Its value is that it accurately reflects our approach to gastronomy, which is to focus not on the whole gastronomic experience (that is the responsibility of the chef), but on what is happening to the food at the molecular level. The problem of producing a perfect boiled egg, for example, is a problem of convincing the string-like albumin protein molecules in the white of the egg to become entangled while leaving similar molecules in the yolk in their native, unentangled state. This is a matter of getting the right amount of heat to the right place. Just how to do this is the central problem, not just of egg-boiling (which was not even on the agenda when we began our 1997 meeting), but of cooking in general.

The transport of heat is a matter of physics, but its rules are so simple that no scientific training is needed to understand them. To work out how the rules apply to practical cooking problems, though, it is necessary to understand how heat affects food flavor and texture, and this in turn means understanding what heat *is*. Unfortunately for ease of communication between chefs and scientists, the true nature of heat is not readily understandable in commonsense terms. It was time for a short history lesson. Luckily, it was one in which food entered in unexpected and even entertaining ways.

Until the middle of the nineteenth century, heat was believed to be an actual fluid. This was a perfectly reasonable, commonsense view, since heat is clearly able to "flow" from hotter to colder places, and it is difficult to imagine this happening unless heat is a real fluid. The fluid even had a name — caloric — and it was believed that "the sensation of heat is caused by particles of caloric passing into our bodies." The commonsense picture of heat as caloric accounted for a lot of the known facts. Addition of caloric to an uncooked piece of meat would, likewise, produce a different material: the cooked version.

Although caloric lived on into the mid-nineteenth century, its death knell was sounded some fifty years earlier by the American adventurer Benjamin Thompson, a man whose personal and scientific lives were both influenced by some unusual encounters with food. When he was in his twenties, and in command of British troops during the American War of Independence, his soldiers used tombstones from a cemetery to build a bread oven. Some of the loaves were distributed to members of the local community, unfortunately with the epitaphs of their dead relatives baked backwards into the crusts. After this "it was considered prudent that he should seek an early opportunity of leaving the country." He moved to England, where his talent for personal advancement proved so great that he became undersecretary of state within four years, and a Fellow of the Royal Society for his research into gunpowder, firearms, and naval signaling.

Moving to mainland Europe, he acquired the title of Count Rumford, Count of the Holy Roman Empire, and became Minister of War for Bavaria. It was in this capacity that he came to be in charge of the Munich arsenal when he made the famous observations that led to his devastating dismissal of caloric. In his own words:

> Being engaged . . . in superintending the boring of cannon, I was struck with the very considerable degree of Heat which a brass gun acquires, in a short time, in being bored; and with the still more intense Heat (much greater than that of boiling water, as I found by experiment) of the metallic chips . . . the source of the Heat generated by friction, in these Experiments, appears to be *inexhaustible*. It is hardly necessary to add, that anything which any *insulated* body . . . can continue to furnish *without limitation*, cannot possibly be a *material substance;* and it appears to me to be extremely difficult, if not quite impossible, to form any distinct idea of anything, capable of being excited and communicated, in the manner that Heat was excited and communicated in these Experiments, except it be MOTION.

Rumford's conception of heat as motion is now commonplace among scientists. We think of the effect of heat on a food, for instance, largely in terms of the increased mobility of the molecules in the food, which consequently become rearranged and disrupted. The long albumin molecules in the white of a boiled egg, for example, which exist as loosely folded balls at room temperature, unfold and wave around as the egg gets hotter, eventually becoming entangled and creating a three-dimensional network which traps the water in the egg white, turning it from liquid to solid and from transparent to opaque. Computer simulations are now available that demonstrate such molecular rearrangements in graphic detail. No amount of picturesque detail, though, will answer a fundamental question. Heat is one thing. Motion seems to be something totally different. How could the two possibly be related?

The solution to this problem required a leap of imagination at least as brilliant as that required for the development of quantum mechanics or the Theory of Relativity. Yet, while everyone has heard of Einstein, few have heard of Julius Mayer, the failed German physician who linked heat and motion through the concept of *energy*.

The full story of Mayer's despairing efforts to get his ideas accepted, culminating in his attempted suicide, is given in Appendix 1 at the end of this book. Suffice it to say here that his ideas were eventually accepted, even though the credit often goes to others, and the notion of "energy" now underpins the whole of science.

What is "energy"? Luckily for ease of communication, the scientist's definition is very close to the way in which we use the word in everyday speech. Put simply, "energy" is anything that can be made to perform physical work, i.e., to move something. The more energy we have, the more we can move, and the further we can move it. A beam of light, for example, can be used to spin a tiny windmill known as a Crooke's radiometer. Light, then, is a form of energy, just as heat, electricity, magnetism, and gravity are also forms of energy, all of which can be used to drive different types of engine. Movement itself is a form of energy, since one moving object may be used to move a second one. Energy of motion has its own name — *kinetic* energy. When we heat food, as Nicholas Kurti pointed out, the increased kinetic energy of the individual food molecules lets them work harder to vibrate, wriggle, and strive to break free from their moorings, eventually undergoing changes that usually make the food more palatable.

The concept of heat as the energy of molecular motion lets us understand many of the events in cooking that would have been a puzzle to believers in caloric. If it is true that caloric + raw food = cooked food, then the addition of caloric at any temperature should eventually cook the food. Yet an egg can be left in water at 50°C for hours without the white setting, while if the temperature is raised to 70°C the white will set within a quarter of an hour; a time that reduces to the classic three minutes or so if the temperature is raised to 100°C, the temperature of boiling water.

This temperature effect — inexplicable using the common-sense caloric picture of old — is easily accounted for using the concept of heat as the energy of molecular motion. The string-like albumin molecules in the egg white have a loosely folded ball structure (technically known as a random coil). This structure is held together by weak attractive forces between those parts of the molecular chain that cross close to each other. The structure is a dynamic one, fluctuating and wobbling as it is bombarded from all sides by surrounding water molecules. As the temperature is increased, the energy of the bombarding molecules correspondingly increases, as does the energy of internal vibration of the albumin chain itself. There comes a point where that energy is sufficient to disrupt the weak linkages that hold the structure together. This happens at a fairly precise temperature (around 68°C). Below that temperature, no amount of cooking will disrupt the structure. Above it, the albumin molecules unfold and become free to entangle themselves with other, similarly unfolded, albumin molecules, creating a new structure — a three-dimensional net.

Just one thing needs to be clarified about this (slightly simplified) picture — the difference between the words "heat" and "temperature." Einstein, writing in 1938, believed that "these concepts are now familiar to everyone," but Einstein was wrong. Most people outside science (and a surprising number inside) would still be hard-pressed to spell out the difference between heat and temperature. We needed to be clear about the distinction in our discussion of cooking. With the concept of energy under our belts, the clarification (presented by Nicholas) took about thirty seconds. The distinction, as Nicholas pointed out, is very simple. *Heat* refers to *total* energy. The *temperature* of a material, on the other hand, is a practical measure of the *average* energy per molecule in the material. In cooking, the total energy that is delivered to the dish being cooked depends on the cooking device. Hotplates and grills deliver heat energy at a rate which is more or less constant for a particular setting, so the amount delivered to the dish depends on the time and the setting. Microwave ovens deliver energy in intermittent bursts of constant power, with the relative "on" and "off" times being determined by the setting. The amount of energy that is actually absorbed by the food depends on how much moisture is present and its position in the spatially uneven microwave field in the oven. The effect of the total amount of heat energy delivered on the temperature depends on how much food there is (the more food, the more molecules), on the type of food, and on how the heat energy is distributed within the food. It is the temperature, rather than the total amount of heat energy added, that determines what happens in cooked food at a molecular level. One (usually minor) effect arises from the fact that more energetic molecules, like more energetic people, need more space, jostling each other aside to get it, which is why materials expand as the temperature increases (e.g., the liquid in a thermometer). As the temperature increases, molecules also change their shapes, move to different places, break apart, and join chemically with other molecules. All of these changes (see Appendix 2) alter the flavor and texture of the food. The aim of cooking is to direct those changes in a gastronomically appropriate manner.

The main problem in cooking is how to achieve the appropriate temperature distribution in the food. There are simple physical laws that can be used to predict the temperature distribution. Our aim at the 1997 Erice meeting was to find out whether these laws work in practice during cooking, or whether some foods might have nasty surprises in store.

The two main processes by which heat energy might be transported within food are *conduction* and *convection*. All materials conduct heat; the difference between "conductors" and "insulators" lies only in the rate at which they conduct heat. Meat, for example, is almost as efficient an insulator as the rubber in a wetsuit, but its low heat conductivity is nevertheless sufficient to permit the center to reach a reasonable temperature during cooking.

If the hot material in a food can move, convection also becomes a possibility. Although the notion now seems familiar (as in convector heaters), it was in fact discovered by Count Rumford little more than two hundred years ago, after another unfortunate encounter with food:

> When dining, I had often observed that some particular dishes retained their heat much longer than others, and that apple pies . . . remained hot for a surprising length of time . . . I never burnt my mouth with them, or saw others meet with the same misfortune, without endeavouring, but in vain, to find out some way of accounting . . . for this surprising phenomenon.

Twelve years later, he had a similar encounter with thick rice soup, which had been brought to him hot but which he had left for an hour. His first spoonful, taken from the top, was cold and unpleasant. The second, taken from deeper down, again burned his mouth. Rumford was still puzzled. His puzzlement was due to the fact that water was believed at the time to be a good conductor of heat. Why, then, did these water-laden dishes not cool down faster? As so often in culinary matters, alcohol eventually supplied the answer. The alcohol was in the huge (4-inch) bulb of a specially constructed thermometer that Rumford had taken to a high temperature during an experiment and then left on a windowsill to cool. To his intense surprise, he saw "the whole mass of liquid in a most rapid motion, running swiftly in two opposite directions, *up,* and *down,* at the same time." Looking more closely, he discovered that "the ascending current occupied the [central] *axis of the tube,* and that it descended by the *sides of the tube."* This process, which Rumford called *convection,* is commonplace in cooking. When water is heated in a saucepan, for example, the heated water at the bottom, which expands and becomes less dense than the colder water above, rises to the top, and is replaced by an inflow of cold water, which is again heated in

turn, so that there is a continual circulation of water carrying
heat to all parts of the saucepan (Figure 2.1).

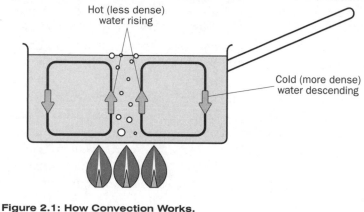

Hot (less dense)
water rising

Cold (more dense)
water descending

Figure 2.1: How Convection Works.
The movement of water in a saucepan.

Convection is vastly more efficient than conduction as a
mode of transporting heat. Water was thought to be a good
conductor only because no one before Rumford had recognized that convection existed. Rumford guessed that water is
really a poor conductor of heat, and that his problems with
apple pies and rice soup had occurred because the free movement of the water was somehow blocked in these dishes. To
check his guess, he deliberately blocked convection in two
pans of hot water, dissolving starch in one and stuffing an ei-
derdown in the second. He found that the water in these pans
cooled much more slowly than did the hot water in a pan to
which nothing had been added to hinder the convection
process. Rumford speculated (correctly) that, in dishes such as
stewed apples and thick rice soup, convection currents are
slowed down or blocked by the presence of fiber and dissolved
substances that are released during cooking. The surface layer
may cool down, but the hot material inside cannot be transported by convection to the surface.

Convection is likely to be similarly blocked in an egg that is
being boiled, since the heat-induced density gradients in the

white are unlikely to be large enough to cause substantial material movement of such a viscous material. Convection is even less likely in dishes like vegetables or roast meat, where the water is held trapped in a matrix of fibers. Conduction, though slow, is likely to be the dominant mode of heat transport in such foods. The disadvantage of this, from the point of view of a chef, is that meat and large vegetable pieces take a relatively long time to cook. The advantage, though, is that these foods (once cooked) retain their heat for a long time. Another advantage is that its basic rules are easy to write down. Those rules, though, are not always the ones that are given in cookbooks or believed by chefs.

Take the simple case of a large, flat slab of meat, such as a steak, cooked in a vertical grill so that it is being heated equally from both sides. If the thickness of the meat is doubled, what does that do to the cooking time? A consensus of chefs (not those at Erice!) got the wrong answer. Many thought that it might not take even twice as long to cook the thicker piece. The correct answer, proved by experiment, is that it will take *four times* as long to cook the thicker piece, if one defines "cooked" as "reaching the same temperature at the center." This is one example of the fact that heat transfer by conduction generally follows a "square rule." To get the heat twice as far takes four times as long.

"Why a square rule?" asked the chefs at Erice. The answer lies in the way that kinetic energy is transferred between molecules in food.

The process starts in cooking when heat energy reaches the food surface, increasing the kinetic energy of the surface molecules. These molecules then pass some of that energy on to their less energetic neighbors by a "knock-on" effect. The energy continues to be passed on to further molecules in relay fashion. The rules that govern this process are statistical, and based on the idea that the energy may be passed in any direction with equal probability, so that the governing equation is the same as that which describes the random diffusion of molecules in a liquid (see chapter one). This equation shows that the time taken for heat energy to travel a given distance by conduction depends on the square of the distance. To travel twice as far takes, on average, four times as long.

The equations for conductive heat transfer were written down by the French mathematician Jean Baptiste Fourier, one of the savants who accompanied Napoleon to Egypt in 1798. The square rule is a solution to Fourier's equation that is accurate for flat pieces of food where the width is very much greater than the thickness. Would it work for a food such as an irregularly shaped roast? There is every reason to expect it to. Solutions of Fourier's equation for shapes other than a flat slab are complicated, but all contain a term in which the time depends on the square of the distance.

Theory, though, is no substitute for experiment, especially where cooking is concerned. We decided at Erice to test the theory with a genuine roast, lovingly prepared by chef Fritz Blank, proprietor of the famous Philadelphia restaurant Deux Cheminées. My task was to lace the roast with fine wire thermocouples, inserted so as to monitor the temperature changes at different depths in the meat. The wires from these thermocouples trailed across the kitchen from the oven to a multichannel recorder, where Fritz and I sat watching while we sipped a reflective glass of wine. Two hours later, the center of the roast had reached Fritz's prescribed value of 45°C, and conference talks were forgotten as the speakers crowded with the rest, eager for a taste. First, though, Fritz insisted that the roast had to be left for forty minutes to "settle." I couldn't understand the reason for this bit of chef's folklore, though I was soon to find out why. The delay gave me an opportunity to keep monitoring the temperature of the roast as it cooled down, while analyzing the data obtained so far. If the square rule held, then a graph of distance squared against time to reach any particular temperature would be a straight line. I tried it for a few different temperatures. When I saw the results, I felt that the glass of wine had been justified. The temperatures in the meat during roasting followed the square rule beautifully.

The roast, meanwhile, had a little surprise in store for us. The thermocouples near the surface showed that the temperature had begun to drop as soon as the roast was removed from the oven. Those nearer the center, though, showed the temperature still *rising!* The center temperature continued to rise for the next forty minutes, eventually reaching 55°C, a temperature appropriate for somewhere between medium and well-done. Does meat, then, disobey the normal rules of conduction?

I quickly realized that the normal rules of conduction were in fact responsible. The cool center of a roast is surrounded by hotter meat, even after the roast has been removed from the oven. The layer of highest temperature will be somewhere between the outside and the middle, and heat will flow from this layer to cooler places, which means that it will flow both to the outside and the inside of the meat. Later analysis showed that the rate at which this process occurs fits very closely with the predictions of Fourier's equation. The analysis also showed that the chef's habit of allowing large roasts to "settle" before bringing them to the table has a very solid scientific foundation. The center of the meat goes on cooking, and the temperature profile also flattens out, so that the meat is more evenly cooked. Meat will also be more evenly cooked if it is roasted for longer at a lower temperature. But how can this be achieved when we want high temperatures to promote the browning reactions at the surface, giving that lovely crispy texture and flavor?

The answer is simple. Start the oven off at a high temperature, then turn the temperature right down after a short time. This is what professional chefs like Fritz do when they are not collaborating in experiments. The square rule still applies, though the actual times are different because of the lower oven temperatures. In fact, the square rule is a good guide for many foods. The differences between cooking times based on the square rule and those calculated from such traditional methods as "20 minutes per pound plus 20 minutes" or "25 minutes per pound plus 25 minutes" are interesting:

**Table 2.1: Calculated Times to Cook a Piece of Roast Beef to
"Rare" Perfection in an Oven at 190˚C (375˚F).**

Weight of Piece (kg)	(lb)	Square rule time (min)*	Mrs. Beeton's time (20 min/lb + 20 min)	My mother's time (25 min/lb + 25 min)
0.1	$1/_4$	30	24	31
0.2	$1/_2$	47	28	36
0.5	1	86	42	53
1	$2^1/_8$	140	64	80
1.5	$3^1/_4$	181	86	108
2	$4^1/_2$	219	108	135

* Adapted from Peter Barham's original calculations in *The Science of Cooking*.

The traditional rule inevitably overestimates cooking times
for smaller pieces of meat and underestimates the times required to cook larger pieces. There will be a crossover point
where the two rules agree exactly for a given weight of meat,
usually the weight with which the writer recommending the
particular rule has had the most cooking experience. The
agreement between the two rules for a range of weights on either side of this crossover point is reasonable. There is an interesting mathematical reason for this range of agreement,
with the main point for the practical cook being that the calculated cooking times diverge more rapidly for weights below
the crossover point than they do for larger weights.

Top-class chefs are very good at estimating cooking times,
and how these times change with weight, without recourse to
the square rule, with which their estimates usually accord
quite closely. The rule for the intelligent domestic cook is:
Practice until you produce the perfect result, and keep a note
of the weight of the portion and the time that produced the result. Then use the square rule to alter the cooking times for
portions with different weights. This sounds very straightforward, but there is a trap. The square rule applies to diameters,
not to weights. To convert from one to the other is tricky unless you are mathematically inclined. In mathematical terms,
the cooking time scales with the square of the diameter but

with the two-thirds power of the weight. The conversion is doable but mathematically complicated, so forget it unless you are a dab hand with a calculator. Instead, just add fifty percent to the cooking time if you double the weight, and proportionately less or more otherwise (e.g., if the weight increase is fifty percent, add twenty-five percent to the cooking time). This simple approximation to the actual rule is surprisingly accurate, as tests with the figures in the table above will quickly show. It's probably the rule that top-class chefs have intuitively worked out for themselves.

The square rule, which applies to so many foods, should surely apply to boiled eggs — and it does. We didn't need to do the experiment at Erice, though. Richard Gardner, Professor of Cell Biology at the University of Oxford, had already done it nine years earlier when trying to understand why his two-year-old son Matthew was able to eat the yolk of a freshly opened boiled egg, but steered clear of the white until it had cooled down. Stimulated by scientific curiosity, Professor Gardner inserted a pair of thermocouples into an egg, one in the white and one in the yolk, and set the egg to boil. Professor Gardner did not interepret his data in terms of the square rule, but we were able to, because he published his results in an extraordinary anthology on food and drink by Fellows and Foreign Members of the Royal Society. The editor of the anthology was (of course!) Nicholas Kurti.

A graph of the results is shown in Figure 2.2. The lower "wiggle" (A) in the curve for the temperature of the white is an artifact, arising from moving the thermocouple after the measurements had started. The upper "wiggle" (B), however, has real meaning. It occurs at the temperature (and time) when the white sets, and setting takes energy, therefore the temperature of the white stays constant (if energy is coming in by conduction as fast as it is being used to rearrange molecules) or even drops (if more energy is required to rearrange the albumin molecules than is available from conduction).

Professor Gardner allowed his egg to cook for thirty minutes, an appropriate procedure for a lover of very hard-boiled

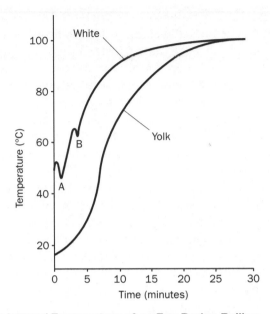

Figure 2.2: Internal Temperature of an Egg During Boiling.
Redrawn from Richard Gardner, "On Boiling Eggs," in Kurti, N. and G. (eds.), *But the Crackling Is Superb.*

eggs, and also for a scientist interested in testing the application of the square rule, which gives an excellent fit to Professor Gardner's data. The production of a perfect soft-boiled breakfast egg, though, requires the cooking to stop after a much shorter time. In fact, if Professor Gardner had removed the egg after three and a half minutes (the time at the end of the second "wiggle") and opened it immediately, his egg would have been perfectly cooked (assuming that the thermocouple in the white had been measuring the temperature at a point very close to the yolk). We know that the egg would have been perfectly soft-boiled because at the temperature when the white sets, the yolk is still runny. This is due to the fact that the protein molecules in the yolk are each wrapped around a tiny core of oil. It takes more energy to release the protein from the oil surface than it does to unwind an albumin molecule in the aqueous environment of the white — the yolk proteins are not free to move around and become entangled until the yolk reaches a higher temperature than the white. The yolk, in fact, only sets above a temperature of 68°C, so the problem of boiling an egg becomes a matter of getting the white above 63°C, while keeping the yolk below 68°C.

For the cook without a thermocouple, it's a matter of delicate timing. The square rule lets us calculate the length of time required, which is astonishingly sensitive to the size of the egg concerned. The calculation was performed by Dr. Charles Williams of Exeter University in 1998. He presented his results in the form of an equation, from which I have calculated the figures below:

Table 2.2: Calculated Times to Boil the Perfect Egg.

Egg diameter at widest lateral point (mm)		Cooking time (min)	
		Initial temp. 20°C	Initial temp. 4°C
small	39	3.34	3.75
small	40	3.5	4.0
medium	42	3.9	4.4
medium	44	4.25	4.8
medium	46	4.6	5.2
large	48	5	5.7
large	50	5.5	6.2

These figures show that James Bond was right — so long as the French Marans hens laid at least some eggs that were 39 millimeters in diameter, and that these eggs were kept at room temperature. Bond, in his fastidious way, would of course have had a metal ring available with a diameter of 39 millimeters, and would only have used eggs that just fit through this ring, so that a cooking time of three and one-third minutes would have been perfect. For medium eggs at room temperature, place directly into boiling water, and allow around four minutes cooking time. The times will shorten if the egg is allowed to "settle" before being opened, since the center will continue cooking even after the egg is removed from the boiling water, just as the center of a roast joint keeps on cooking after it has been removed from the oven. Such a procedure will produce a more delicately textured egg, with the white not quite so rubbery, since fewer cross-links will have been formed between the albumin molecules. A sophisticated way of tackling this problem, devised by Fritz Blank and used in his restaurant, is to cook the eggs for a shorter time than normal and then to roll them in crushed ice while the inside goes on cooking, so that the residual internal heat goes towards cooking the center but does not overcook the outside of the white and create a rubbery texture. However, as Nicholas pointed out, there is an even better way. His approach, later elaborated by Hervé This in a letter to the magazine *New Scientist*, was based on the knowledge that the white sets at a lower temperature than the yolk. All that is needed, then, is to boil the egg in a liquid whose boiling point is between the two setting temperatures. The white will eventually set, but the yolk never will. And the egg can be boiled for as long as the cook likes.

Someone with the resources of a scientific laboratory can achieve the appropriate temperature (between 63°C and 68°C) by boiling the water under reduced pressure. This requires elaborate (and expensive) apparatus, together with appropriate safety measures. The alternative is to use a different liquid altogether, one with a boiling point of 64–66°C. There are a few such liquids. One, common in chemical laboratories, is methanol (also known as wood alcohol), which has a boiling point of 64.6°C. There are only three problems. The first is the flavor that the methanol is likely to impart to the egg through its porous shell. The second is availability — wood alcohol is available to chemists for scientific purposes, but its commercial use is mostly as a poisonous adulterant in methylated spirits. It is the third problem, though, that presents the most difficulty. Methanol vapor is highly inflammable, and liable to catch fire even on an electric hotplate. The conclusion is that there is an ideal, scientific, guaranteed method to boil the perfect egg, but don't, whatever you do, try it at home. James Bond might have been able to get away with it, dousing the resulting flames with one of Q's special gadgets that he doubtless carried in the trunk of his Bentley. He may even have been able to use the flaming egg as a Molotov cocktail. For a proper gastronomic experience, though, the rest of us will do far better by sticking to a combination of water and simple arithmetic.

3
the tao of tools

One of my greatest problems as a student learning about science was wanting to understand the logical basis of the ideas that were being presented to me. It sounds like just the sort of problem that a scientist ought to have, but many of the most fundamental ideas were simply presented to us as facts to be accepted and used. My more successful contemporaries accepted this approach, got on with things, and in due course became professors and even vice chancellors. I, in the meantime, spent endless hours trying to figure out where things like the Schrödinger equation (which governs all of quantum mechanics) came from, or how the concept of energy arose.

I eventually found out that the Schrödinger equation was a complete guess, perhaps the most brilliant guess in the history of science, and that the concept of energy grew gradually from efforts to understand what heat was. When I looked more closely, though, I found that energy was always defined in terms of its ability to perform physical work, so work, therefore, was an even more fundamental concept than energy. Where did this concept come from? It took me thirty-five years to find out. When I did, the answer came as a complete shock. The definition of "work," the most fundamental quantity in science, came entirely from intuition. We don't even know whose intuition. We do know, though, that the definition can be used to figure out how best to use the tools that supposedly save us work. They don't, of course. What tools do is to make work possible, by changing the balance between the force we need to exert and the distance through which we have to move the point of application of that force to do a given job. This chapter describes how that principle arose, and how people from Archimedes onwards have used it to achieve objectives ranging from lifting a Roman galley out of the water to removing recalcitrant nails from hardwood. The same principle can be used to work out the most efficient way to use tools. As the reader will find, this is not always the way that they are usually used by handymen and tradesmen.

Is it best to drive a nail into a piece of wood with a series of light blows rather than a few heavier blows? When removing a nail with a claw hammer, does it help to place a small block of wood under the head of the hammer? Are long screwdrivers easier to use than shorter screwdrivers with the same blade size? How sharp does a chisel really need to be? From practical experience, the answers to these questions are: yes, yes, yes, and very. There is a right and a wrong way of using tools; an effective and an ineffective way; a way that makes the work easier and a way that makes work harder.

Practical experience is codified in books such as Morgan's classic *Woodworking Tools and How to Use Them*, a book that I devoured eagerly when young, partly in the hope of showing my father that some aspect of his workshop teaching had been wrong. Sadly for my youthful hopes, the rules were just as he had said (I later found that he too had read Morgan's book). Nowhere, though, did the books say *why* the rules were as they were. In this they differed from the other books that were shaping my future life — the popular science books that addressed the question of *why* instead of *how*. The world portrayed in those books seemed to me a more important one, dealing with things that really mattered, and far removed from the mundanely practical. I did not realize how the two worlds feed off each other, and how they are both part of one larger, interconnected world of understanding.

I was not the first to make such a mistake. Archimedes, born nearly three hundred years before Christ, shared my misconception that practical things are much less important than ideas. Even though he invented many practical devices, he seldom thought them sufficiently important to write down their details for posterity. It is because of this attitude that he left no description of the first hand tool to be designed from scientific principles.

As hand tools go, it was rather large; large enough, in fact, to lift a Roman galley clear of the water and shake the frightened soldiers out like weevils from a sailor's biscuit. Despite its size, though, it still fitted the Oxford Dictionary definition of a tool as "a mechanical implement for working upon something . . . held in and operated directly by the hand, but also including some simple machines." This particular one probably required quite a few hands to operate. We know from historical records that it took the form of an asymmetric lever, mounted on the seawall of Archimedes' home town of Syracuse as protection against an invading Roman fleet. The lever arm must have been like a long tree trunk. The short end hung over the sea, with a claw-like grab suspended from it. Once the grab had hooked into some part of an attacking ship, teams of men or animals pulling down on the long end of the lever could lift the ship clear of the water. The device was so effective, according to the Greek historian Plutarch, that if attacking Romans saw a piece of wood projecting over the seawall it was enough to make them turn tail and head for the open sea (Figure 3.1).

Archimedes' great machine worked first time because he understood not just *how* levers work (the ancient Egyptians knew that) but *why* they function in the way that they do. He had worked out the law of the lever years earlier, and described it in a book entitled *On Balances or Levers,* a book that is unfortunately now lost. Put mathematically, as Archimedes would have done, the law simply states that the product of the load and the length of the lever arm is the same for both loads. An earlier writer, mysteriously called the pseudo-Aristotle, worked out the law intuitively and put it in a more understandable way: ". . . as the weight moved is to the weight moving it, so, inversely, is the length of the arm bearing the weight to the length of the lever arm nearer the power."

I discovered the lever principle experimentally, as many children do, when I found that I was able to lift my father's weight on a seesaw, provided that he sat sufficiently close to

**Figure 3.1: Hypothetical Reconstruction of Archimedes'
Ship-Lifting Lever.**

the middle while I sat on one end. He weighed three times as
much as I did, so I had to sit three times as far away from the
middle (Figure 3.2).

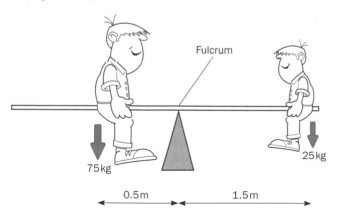

Figure 3.2: Balancing on a Seesaw.
The fulcrum, or balance point, is the point about which rotation occurs. Technically,
a seesaw is a *first-order* lever, since the loads are on opposite sides of the fulcrum.

I quickly learned to apply the lever principle to other situations, including the memorable occasion when I used one of my father's carefully sharpened chisels to jimmy open the tightly stuck lid of a tin of candies. My father said that he would teach me to use his chisels in that way, but I already knew, and avoided the painful lesson by hiding in the wardrobe.

As time went on I learned that there are other forms of lever, such as the wheelbarrow. My struggling attempts at age twelve to use one to shift a load of wet cement led a friend's mother to scream in alarm: "Oh, Lenny, don't do that; you'll bust a kafoops valve!" The lever principle worked fine, though, and I didn't bust a kafoops valve, whatever that is (Figure 3.3).

It wasn't until I reached high school that I learned the very simple mathematics underlying the use of levers, and realized that this could be used to calculate in advance, as Archimedes had, just where to put a fulcrum and how to design a lever to do a given job. I now use this knowledge in the design and construction of scientific measuring instruments for my own

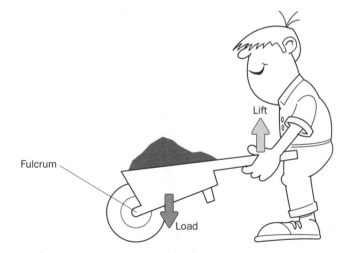

Figure 3.3: The Wheelbarrow.
Technically, a wheelbarrow is a *second-order* lever, since the load and the lifting force are on the same side of the fulcrum, with the lifting force being further away than the load.

and other people's research. Only recently, though, did I start to wonder whether I could apply this and a few other equally simple physical principles to improve my efficiency in using hand tools to tackle simple domestic jobs.

The stimulus for the thought was a conversation over early-morning coffee in the Physics Department at Bristol University, where I raised the question of why long-shafted screwdrivers appear to be easier to use than shorter screwdrivers of the same blade width. It's a typical scientist's "why?" question. Contrary to popular belief, there is no prescribed, rigorous, logical path for finding the "scientific" answer to such a question. It's very much a matter of style, with individual scientists following one of three broad approaches. The first is to work out an answer based on fundamental principles, not accepting or even caring what others might have said before. Some of the scientists best-known to the public, such as Archimedes, Newton, and Einstein, worked in this way. Those who are capable of using this approach tend to be the gadflies of the scientific community, respected, and sometimes feared, by their fellow scientists.

One such was John Conrad Jaeger, the Australian coauthor of an influential applied mathematics textbook. A friend of mine was present when Jaeger was in the audience for a seminar given by a student just finishing his Ph.D at the University of Tasmania. It was in the early days of computers, and the student was reporting his success in using one to tackle a particularly complicated problem. My friend describes how the student ended his seminar with a flourish as he produced the complete solution to the problem, then stood back with his supervisor, basking in the adulation. Jaeger asked if he might come to the blackboard, which was still covered with the student's equations. Picking up a piece of chalk, he began mumbling to himself, "The limit of this function is so-and-so; that expression is approximated as this; these two terms cancel . . ." After a minute or so, he produced the solution to the problem, glanced over at the student's results, said, "Yes, that's right," and sat down.

Science could hardly survive without people like Jaeger, who have the capacity to see straight through to the heart of a problem. I have been privileged to work with a few such people in the course of my scientific career. One was a mathematician who could solve problems where I wouldn't even know how to start. I could at least understand the answers, since I had once done a degree in pure mathematics. I incautiously revealed this fact one day, whereupon he turned, looked at me with astonishment, and said, "You?!" At Bristol they were more polite, but equally forthright, when I raised the problem of the screwdriver. My argument was that longer screwdrivers must be easier to use because they can be tilted slightly without the head coming out of the slot, thus providing added leverage. My colleague Jeff Odell calculated, in the time that it took for one sip of coffee, that a tilt of a few degrees would add little to the rotational force, and commented that, in any case, leverage had little to do with it. "A firmly held screwdriver," he commented, "is simply an extension of the arm. The ease of rotating the forearm is the only thing that counts. Longer screwdrivers are probably easier to use because the handles are bigger and easier to grip without slipping."

I had no answer to that other than to go away and try the experiment. The results of this approach are given later in this chapter. It's the approach that I have most often used in my scientific career, and one that puts me firmly in the second camp of scientists — those whose first instinct is to measure something and think about the results afterwards.

My favorite example of this style of tackling scientific questions is that of Lord Rutherford, the bluff New Zealander who dominated British physics early in the last century. Rutherford was studying alpha-particles — heavy, positively charged particles that are emitted from some radioactive materials like bullets and which travel at an appreciable fraction of the speed of light. He later recalled that "One day Geiger [his right-hand man, and the inventor of the Geiger counter] came to me and said, 'Don't you think young Marsden whom I am training in radioactive methods ought to begin a small research?' Now I had thought so too, so I said, 'Why not let him see if any alpha-particles can be scattered through a large angle [from a thin piece of gold foil]?'"

On the face of it, this was a crazy experiment, with no realistic chance of success. At the time, atoms were believed to be like tiny individual plum puddings, with negatively charged electrons (the raisins) dotted randomly about in a spherical haze of positive charge (the pudding). Speeding alpha-particles should pass straight through a material built from such atoms with no more trouble than a rifle bullet through a real plum pudding. Yet when Marsden tried the experiment, many alpha-particles were deviated widely from their paths, and some were even reflected back in the direction from which they had come. According to Rutherford, this was "quite the most incredible event that has ever happened to me in my life . . . It was almost as incredible as if you fired a fifteen-inch shell at a piece of tissue paper and it came back and hit you."

Rutherford had no reason to expect such a spectacular effect, but he had done what many scientists with a good experimental instinct do; that is, to try something that could be important *if it worked*, and not bothering too much about whether it was actually likely to work. The less likely an experiment is to work, the more significant the result is likely to be. In this case, the result was very significant, and led Rutherford to develop the modern picture of the atom, with the "haze of positive charge" actually concentrated as a central lump (the atomic nucleus), substantial enough to deflect alpha-particles that approach too closely.

Rutherford's approach of *trying out* an instinctive idea was probably the one used by the early inventors of hand tools, including the unknown person who first thought of tying a handle to the piece of rock used to smash animal bones, thus turning that piece of rock into a hammer. The real Rutherford of the hammer world, though, was the person who, after several thousand years of poorly tied heads flying off hammer handles, conceived the idea of putting a hole in the head and fitting the handle into that.

Many varieties of hammer have since been developed. I was interested to find out whether anyone had taken Rutherford's next step, the one that marks a true scientist, of asking *why* hammers, screwdrivers, and other hand tools work in the way that they do and, if so, whether this information had been or could be used to optimize the way that we now design and use these tools. To do so, I enrolled myself in the third and largest scientific camp — that composed of scientists who, sensibly, look to find what others have done before they go on to consider a question further themselves.

Some of the very best scientists belong in this camp, providing the mortar that holds the whole structure of science together. They are people like my early collaborator Jacob Israelachvili (now at Santa Barbara), the scientist who first measured the forces between surfaces so closely spaced that no more than a few atoms could have fitted between them — an experiment that many, including myself, had thought impossible. He performed these experiments for his Ph.D, and later told me that he had spent two and three-quarter years out of the allotted three in studying what others had done and using this knowledge to design and build equipment to do the job better. He then took just two months to perform the measurements that made him famous, but his success was due in large part to the foreknowledge with which he had armed himself.

The design of Jacob's equipment was based on an intimate understanding of the scientific principles involved. Is the same true for the development of hand tools? I went in search of the written evidence and quickly found, with the help of a friendly engineering librarian, that comparatively little has been written about the science underpinning the use of hand tools. Only a few scattered references bore relation to the subject. Even the closely printed thirty-two-page article entitled "Tool" in the famous 1911 edition of the *Encylopaedia Britannica* (an article still referred to by the modern edition) failed to say a single word on *why* tools are designed and used as they are. I sought out Stuart Burgess, a design expert who runs a second-year course on machine tools at Bristol University.

Could he help? He was only too pleased. Not to help, but to find that I was writing something on the subject. There was nothing of this nature available, he said, and promised to recommend my book to his students when it was finished. Perhaps, I thought, hand tools would feature in physics courses as exemplifiers of simple mechanical principles. I asked my physicist friends and found that, while the basic principles are taught, such applications are not. I was on my own.

Well, not entirely on my own. A lifetime in science has given me plenty of people to talk to and with whom to try out ideas. That's how science usually works — not by people sitting alone in ivory towers, but by people sharing ideas, talking about results, and suggesting new approaches in environments that are sometimes far removed from the traditional lab. Science, in other words, is a community activity, and I have more than my fair share of friends and colleagues who are willing to offer help, ideas, and criticism, in varying proportions. It was to these people, and to my handyman and tradesman friends, that I turned in my quest to understand the scientific *tao* of tools.

My first question was "Why do we use tools?" Galileo berated tradesmen four hundred years ago for holding the mistaken belief that we use tools because they make jobs easier. Despite Galileo's strictures, I suspected that the same belief would still be current today. I was wrong. Every one of my practical friends gave me the scientist's answer, which is that tools don't so much make jobs easier as make jobs *possible*, by reducing the brute force required to manageable proportions. An example is the use of a car jack. A person using an average car jack can lift a 500-kilogram car with a force that would only lift 5 kilograms directly.[1] The weight lifted is a hundred times the force applied, and so the jack is said to give a

[1] Scientists are used to thinking of force in terms of *Newtons*. To lift 5 kilograms would take a force of 49 Newtons, a quantity that most people outside science would find hard to visualize. To make life easier for the reader, I will henceforth refer to forces in terms of the mass that they would lift. When I say, for example, "a force of 5 kilograms," I mean "a force that would lift 5 kilograms at the Earth's surface" (the same force would lift more on the Moon), with a brief nod of apology toward those who would prefer more exactitude of nomenclature.

mechanical advantage of 100:1. When I used my childhood weight
of 25 kilograms to lift my 75-kilogram father from the ground on
a seesaw, I was exerting a mechanical advantage of 3:1.

The downside of a mechanical advantage is that the reduc-
tion in force must be paid for by an increase in the distance
over which the point of application of the force is moved.
Whatever height the car is raised to, the jack handle must be
pumped through a hundred times that distance. There's no escape. Why is there no escape? The reason was discovered by
Galileo, who used a series of very clever arguments to show
that, in the performance of any job requiring the application
of force, *force multiplied by the distance through which its point of
application is moved* is a constant quantity that can't be changed,
no matter how much we wriggle or how cleverly we design
our tool or machine.

Scientists now call *force × distance* by another name — *work.*
It is a basic principle of science, derived these days from the
notion that energy can neither be created nor destroyed, that
the work that we have to do in performing a job is unaffected
by the way in which we do the job.

Galileo knew nothing of the principle of the conservation of
energy, and I had not even seen his approach to the question
until I began putting together the material for this book.
When I eventually did find his argument, written for a lay audience, I wished that I could write like that. It was so auda-
ciously simple that I burst out laughing while reading it. In his
own words: ". . . the advantage acquired from the length of the
lever is nothing but the ability to move *all at once* [my italics]
that heavy body which could be conducted only in pieces by
the same force . . . and with equal motion, without the benefit
of the lever." In other words, if we cut our 500-kilogram car
into a hundred equal pieces, each weighing 5 kilograms, and
then lifted each of these pieces through 30 centimeters by
hand, we would be doing the same work (5 kg × 30 cm × 100)
as if we had pumped the jack handle through 30 meters to lift
the whole car at once (where the work would be 500 kg × 30

cm). All that the jack has done is to let us change the balance between force and distance. The product of the two remains the same. Galileo went on to show that the same argument holds for any tool.

Galileo's principle that a tool can't change the actual amount of work needed to do a job, but only change the balance between force and distance, holds good so long as all of the work goes into doing the job. If doing the job without a tool involves wasting some of the work, then the use of a tool obviously does save work. If, for example, we try to slide a load of bricks along a path by hand, then a lot of the work that we do goes into overcoming friction between the bricks and the path rather than moving the bricks. The extra work is turned into heat energy that is then dissipated into the surroundings and cannot be recovered. If we carry the same load of bricks along the path in a wheelbarrow, we avoid the friction, and save the extra work.

It follows from the discussion above that there are two questions to be answered when it comes to the effective use of hand tools:

1. Is the mechanical advantage as high as it could or should be?
2. Is the waste of energy (e.g., in friction) as low as it could be?

I decided to ask these questions of some of the common hand tools used by tradesmen, handymen, and do-it-yourselfers. Even in the limited time available, I ended up with enough material to make a whole book in its own right. What follows is a selection of the points that I found most interesting and which handymen and do-it-yourselfers might find most useful.

In order to categorize the tools in some way, I turned again to the article "Tool" in the authoritative *Encyclopaedia Britannica* of 1911. According to the writer, the Victorian authority Joseph G. Horner, tools fall into just five groups:

    I   Chisels
   II   The shearing group (e.g., scissors)
  III   Scrapers
   IV   The percussive group (e.g., hammers)
    V   The molding group (e.g., trowels)

There was no mention of tools based on the lever principle, and no mention of another important group: tools that use a wedge action. At the risk of incurring the wrath of Mr. Horner's ghost, I decided to examine just one of his categories, but to add two of my own. My full list is thus:

1. Tools based on the lever principle
2. Tools that use the wedge principle
3. Percussive tools

Tools Based on the Lever Principle

The Claw Hammer

According to published tables, the initial force needed to withdraw a 2-inch (50-millimeter) nail driven into the side grain of a block of seasoned hardwood is equivalent to lifting a weight of 26 kilograms. That is why it is so difficult to pull a nail out with a pair of pliers, and why we use a claw hammer as a lever to do the job.

The design of a typical domestic claw hammer makes it very easy to start the pull. The mechanical advantage is enormous, since the fulcrum is very close to the nail. My own 700-gram claw hammer has a handle length of 330 millimeters, with the initial contact point of the hammer head with the wood being only 10 millimeters from the nail axis. A mechanical advantage of 330/10 = 33 means that I only have to apply a force of less than a kilogram to get the nail started (Figure 3.4).

The mechanical advantage drops very fast as the nail starts

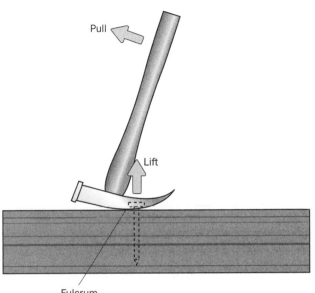

Figure 3.4: Claw Hammer Used as a Lever to Remove a Nail.

to lift, because the contact point of the head with the wood moves further away from the nail as the hammer rocks on its curved spurs. Eventually, the edge of the hammer face, 110 millimeters away from the nail for my hammer, contacts the wood, by which stage the mechanical advantage has dropped to a mere three and the nail has been lifted only 20 millimeters. If the hammer is rocked any further, the head will make a dent in the wood. The force that was originally pulling the nail straight up is also now pulling it at an angle, with the potential for further damage.

Is this the best that can be done? Professional tradesmen sometimes improve the situation by placing a small block of wood under the hammer head. My father used to explain that this was to prevent damage to the work, an advantage that has to be paid for by loss of the initial mechanical advantage, since the fulcrum is now further from the nail at the start of the pull (Figure 3.5).

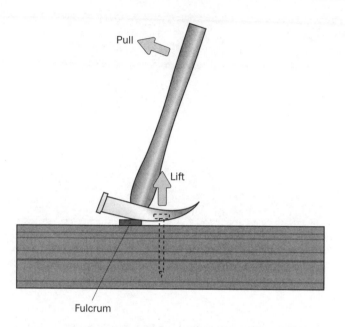

Figure 3.5: Use of a Small Block of Wood to Improve the Range of Movement of a Claw Hammer.

The mechanical advantage is still quite good, however, and also does not change very much through the whole lifting process. If the piece of wood is a 10-millimeter cube, for example, and placed near the base of the handle, the mechanical advantage with my own hammer drops from 10:1 to 8:1 during a single pull, which means that I have to exert an initial force of 2.6 kilograms, a value that is still satisfactorily low. Furthermore, the job gets easier as the hammer is rocked, even though the mechanical advantage is dropping. This is because the pulling force that is needed is proportional to the length of embedded nail; with half the nail withdrawn, only half the force is needed. The real advantage of the little block of wood, though, is that the nail can be lifted through a full 35 millimeters in a single pull. With only 15 millimeters left embedded, the job can now be finished off by a direct pull with a pair of pliers, since this now requires a force of only $26 \times (15/50) = 7.8$ kg.

In thinking all of this through, I came up with another idea, which I will claim as my own until someone comes along and tells me that tradesmen have been doing it for ages. Make a stepped block out of several pieces of wood, and slip it progressively further under the hammer head as the nail is lifted. The stepped block can be kept as part of a regular tool kit, and enables nails of any length to be withdrawn in a single operation.

The Wrench

The ideal wrench is one that exerts pure torque — in other words, all the force exerted goes into rotating the bolt or the nut. My ill-remembered courses in applied mathematics taught me that this is only possible if the wrench is symmetrical, as in the design shown in Figure 3.6. An advantage of such a design for a hand wrench is that both hands can be used to exert force.

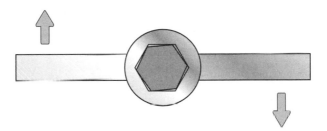

Figure 3.6: Torque Generated by a Two-Handled Wrench.

The forces on the two ends are equal, but act in opposite directions. If two such forces lie in the same line, as they would, for example, if generated by two equally strong individuals pulling on a rope, then they would be in balance and nothing would move. Separated laterally in space, though, the forces constitute a *couple*. The torque, or twisting force, exerted by a couple is simply the force multiplied by the separation distance (my school physics teacher, well attuned to the preoccupations of the adolescent mind, taught us that "the closer the couple, the less the torque").

It is not so easy to work out the mechanical advantage that the wrench above would convey over turning the bolt with the fingers, because the grip and the actual movement of the muscles are quite different in using the fingers or grasping two ends of a bar and pulling on one end while pushing on the other. A better comparison is with wrapping a couple of pieces of tape, one above the other, around the bolt head in opposite directions and pulling on the free ends (quite a good trick, incidentally, if no wrench is handy). In this case, the same force can be exerted as in using the wrench. The mechanical advantage from using the wrench is just the ratio of the torques in the two cases. Since the forces are equal, the mechanical advantage is simply the ratio of the distances by which the forces are separated laterally in the two cases, i.e., the ratio of the overall length of the wrench to the diameter of the bolt head.

Unfortunately for the peace of mind of calculating physicists, most wrenches have only one handle, and working out the mechanical advantage is not so easy. A single-handled wrench does not exert pure torque — it also exerts a lateral force which tends to move an unsecured job sideways. If the job is secured by a clamp or other means, the sideways force is still there, pushing on the bolt or nut and increasing the frictional resistance that has to be overcome, thus decreasing the real mechanical advantage. For the purposes of a simple analysis, a single-handled wrench can be viewed as a lever, where the nearest point of contact on the nut or bolt head is the fulcrum, and the furthest point of contact transmits the force to turn the nut about that fulcrum (Figure 3.7).

This picture is oversimplified (since the fulcrum can move), but it does let us calculate the mechanical advantage for such a wrench, which is simply the distance from the end of the handle to the point of first contact with the nut or bolt head divided by the length of one side of the nut or bolt head. In the example above, the mechanical advantage is 25/5 = 5:1. This

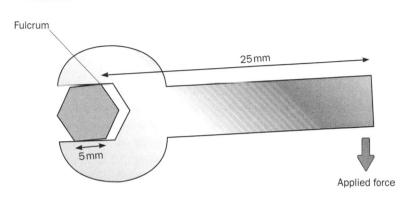

Figure 3.7: Lever Action of a Single-Handled Wrench.

is half the mechanical advantage that a similar wrench with two handles would provide.

An easier way of thinking about the difference between one-handled and two-handled wrenches, and one that is very representative of how physicists think, is to fall back on Galileo's principle that the total work done is not affected by how it is done. This means that *force* × *distance* must be the same in the two cases. With the two-handled wrench, the two points of application move a total of twice as far as the single point of application for the one-handled wrench. Since the distance is doubled, the force must be halved, which means that the two-handled wrench is twice as efficient, i.e., it has twice the mechanical advantage of the one-handled wrench. Even so, a single-handled wrench is a pretty effective weapon. But why do we need such a weapon? What is the actual advantage of tightening a nut or bolt using a wrench? The reason, as I found out in practice when designing high-precision equipment, is that the use of a wrench lets us apply enough force to actually stretch the bolt. The stretched bolt acts as a very strong spring, generating a force that increases the friction between the mating male and female threads, and which also increases the friction between the head and the nut (if there is one) with the corresponding surfaces of the work, making it more difficult for the bolt to work loose. The amount by which a bolt needs to be stretched in order to stay tight depends very much on its environment. The bolts in a vibrating car engine, for example, obviously need to be done up more tightly than those holding a wooden bench together. Bolts holding metal pieces together can also be done up more tightly than those passing through wood without risk of deforming or damaging the work. But how tightly? How is the torque that is used to do up a bolt related to its degree of stretching and to the spring force that it exerts? What spring force should one aim for? Are there simple rules that the handyman can follow?

I searched for the answers in engineering reference books, and rapidly found myself immersed in a morass of formulae which described the effects of bolt diameter, bolt material, pitch, shape and depth of the thread, and even whether the bolt is likely to be subjected to extra forces after it is done up (e.g., bolts holding the head onto a car engine block). Typical of the formulae that confronted me was that of the minimum length of engagement between two mating threads to avoid stripping the external thread before the bolt actually breaks:

$$\frac{2 \times \text{(tensile stress area of screw thread)}}{3.1416 \times \text{(internal thread dia.)} \, [0.5 + 0.5773 \times \text{(threads per inch)} \times \text{(external thread diameter − internal thread diameter)}]}$$

I find it depressing enough when I am forced to use such formulae in a professional context, and would be the last person to inflict them on a handyman. I looked to see whether there might be a better way.

There was. Hidden among the formulae and tables was the information that "experiments made at Cornell University [on behalf of the car industry] . . . showed that experienced machinists tighten nuts with a pull roughly proportional to the bolt diameter," and that "the stress due to nut tightening was often sufficient to break a half-inch bolt, but not larger sizes." As a result of this study, engineering practice was changed —

not to reduce the force applied to the bolts, but to use larger bolts that the mechanics couldn't break!

It is possible, though, to use the information in a different way. The tensile strength of the bolts in question was around 10,000 kilograms (in other words, a rod made of the bolt material could support a weight of 10,000 kilograms). A box wrench designed to tighten such a bolt typically has a length of 20 centimeters (0.2 meters), and by experiment I have found that a reasonably strong man can pull on the end of such a wrench with a maximum force of around 30 kilograms. A torque of 30 kg $\times$ 0.2 m (= 15 kg.m) is thus sufficient to break a half-inch (12-millimeter) steel bolt. Bolts made from weaker materials such as brass or aluminum will require a correspondingly lower torque to generate the same breaking tension. A practical compromise to fasten a half-inch bolt as tightly as possible without the risk of breakage is to use a torque of no more than 8 kg.m. The limiting torque depends on the cross-sectional area of the bolt. The limit for one-inch (25-millimeter) bolts, according to the criterion above, is thus 32 kg.m, while that for quarter-inch (6-millimeter) bolts is 2 kg.m. This means that the "experienced machinists" got it wrong — to generate the same tension in bolts of different diameters, they should have been pulling with a force proportional, not to the diameter, but to the square of the diameter.

If you know the maximum force that you can exert one-handed (this can be estimated by finding the heaviest weight that can be lifted one-handed), you can use this information to work out where to grip a wrench so as to produce the maximum safe torque. Table 3.1 provides some sample values, with the assumption that the longest wrench available is 250 millimeters in length.

Undoing a bolt is a different story. It takes more force to undo even a clean, well-oiled bolt than it does to do one up, because the initial force to overcome friction and get the mating surfaces sliding (head and nut surfaces against the work, and two threads against each other) is greater than that required to keep them sliding, so wrenches will need to be held a little further

Table 3.1: Where to Grip a Wrench.

Maximum force that operator can generate (kg)	Bolt diameter (mm)	Distance of grip from bolt head (mm)
10	4	88
	6	200
	8	Full length
	10	Full length
	12	Full length
20	4	44
	6	100
	8	170
	10	250
	12	Full length
30	4	30
	6	67
	8	113
	10	170
	12	Full length
40	4	22
	6	50
	8	85
	10	125
	12	200

out. When it comes to undoing a rusted bolt, the problem is not that the two threads are "stuck together" by the rust. The real problem is that, as iron turns to rust (a complex reaction product of iron, oxygen, and water) it expands, generating enormous pressures that increase the frictional forces between the threads. One can see how high the pressures can be by looking at a stone into which iron spikes have been driven. As the iron rusts, it is not uncommon for the pressures generated to be so high that the stone is split. Oil is of little use in reducing the frictional forces caused by rust. A better trick (if you have the time) is to use a weak acid such as vinegar to gradually penetrate and dissolve the rust. Alternatively, if the joint is accessible to heat, application of a propane torch will expand the bolt, the nut, and the gap in between to relieve some of the pressure.

The Wheelbarrow

Barrows did not always have wheels. Before the fourteenth century, a European barrow consisted of "a flat rectangular frame of transverse bars, having shafts or 'trams' before and behind, by which it is carried [by two or more men]." Only after A.D. 1300 did the idea of putting a wheel between the front shafts, invented a thousand years earlier in China, finally make its way to the West and permit the contrivance to be operated by just one man. This was one of those inconspicuous advances in technology that helped to make possible such conspicuous markers of progress as the great Gothic cathedrals of Europe. With it, a single worker could now shift large blocks of stone. When some unknown genius put a box between the shafts instead of a flat tray, building rubble and mortar could also be carried.

The load that a single person can shift using a wheelbarrow depends on the position of the wheel. In technical terms, a wheelbarrow is a second-order lever, with the load between the fulcrum (the axle) and the point of application of the force (the handles). If the axle were placed directly under the load, as it was in flat-trayed medieval barrows designed to carry piles of bodies during the plague years, the mechanical advantage would theoretically be infinite, and a single person could move any load at all so long as the barrow did not collapse. Such barrows, which survive in some places as mortuary trolleys, had a wheel on either side, rather than a single wheel in the center. That single wheel makes the barrow more maneuverable, and is essential if the barrow is to be wheeled along narrow planks or manipulated in tight corners. But why don't modern wheelbarrow designers put it in the middle, directly under the load, instead of at one end? The answer lies in stability. So as long as the downward force through the center of gravity of the load does not move outside the triangle defined by the operator's two hands and the point where the wheel contacts the ground, the wheelbarrow will not tip. The reason for this is that the downward force of the load, and the total

Center of Load Inside Triangle –
Wheelbarrow Stable

Center of Load Outside Triangle
– Wheelbarrow Tips

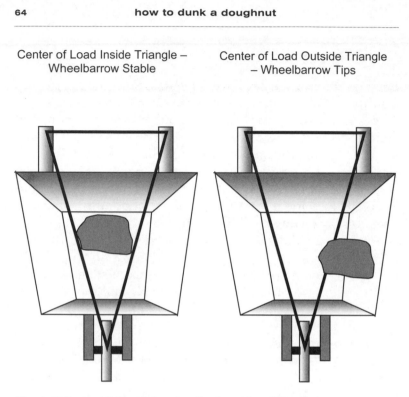

Figure 3.8a and 3.8b: Balancing the Load in a Wheelbarrow.

upward force provided by the operator's hand and the reaction force of the ground on the wheel, constitute a couple that tends to rotate the barrow back in the direction from which it came. If the center of gravity of the load moves outside a line between the operator's hand and the wheel, the situation becomes unstable, since the couple is now tending to rotate the barrow further still. To avert disaster, the operator must exert a counter-couple by lifting on the near handle and pushing down on the far handle. The action must be fast, and the further the barrow tilts the more difficult it becomes, as I found early on as a twelve-year-old.

The problem with putting the wheel in the center of the barrow, with the load directly above, is that the situation is always unstable, since the load is now at the apex of the imaginary triangle formed by the hands and the wheel, and *any* load shift or wheelbarrow tilt is sufficient to take the center of gravity outside that imaginary triangle. The only way to get around this situation is to construct the wheelbarrow in such a way that the center of gravity of the load is below the level of the point of contact of the wheel with the ground. Then the couple created when the wheelbarrow inevitably tilts will act to restore the original position rather than to tilt the barrow further.

It may seem impossible to make such a wheelbarrow, but I have managed to design one that should work when wheeled along a plank, with two equal loads hanging down on either side of the plank. My design (patent not pending) is shown in cross-section (viewed from the front) in Figure 3.9.

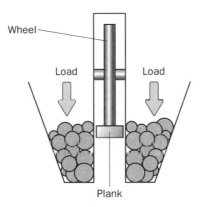

Figure 3.9: Design for a Wheelbarrow that Can Never Tip Over.

For the moment, though, I am stuck with using an ordinary wheelbarrow. Is it possible to use such a barrow more efficiently? It's a compromise between mechanical advantage and stability, which really depends on the load that the user wants to shift. For maximum mechanical advantage, the center of gravity of the load should be as close to the front of the barrow as possible. For maximum stability, the load should be as far back as possible (so that the center of gravity stays within the imaginary triangle even when the wheelbarrow is badly tilted), and spread so as to keep the center of gravity as low as possible.

In the end, it's up to the user. The only extra tip that I have received from my tradesman friends is to keep a couple of lengths of hollow pipe handy that can be slipped over the handles to increase their effective length, and hence increase the mechanical advantage, if an extra-heavy load needs to be moved.

Tools that Use the Wedge Principle

Wedges

The wedge is a very ancient tool. As with other tools, its action is based on a trade-off between force and distance. The action is simple to understand by visualizing the wedge being used to lift something, such as a slab of rock. The mechanical advantage is simply the ratio of the distance that the wedge has to be driven divided by the height through which the rock is lifted.

In the example shown in Figure 3.10, the mechanical advantage is 6:1. The wedge thus allows the rock to be lifted with one-sixth of the force that it would take to do the job directly, at the expense of having to move the point of application of that force (i.e., the base of the wedge) six times as far.

So far, so simple. The real advantage of a wedge, though, is that it lets us change the direction of the force. We drive the wedge sideways, but the rock is lifted upward (the same thing happens when we force a claw hammer under the head of a

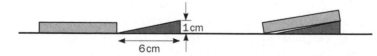

Figure 3.10: Mechanical Advantage of a Wedge.

nail). The mathematical reason for this is explained later, in chapter six. All the information that is needed here is that the mechanical advantage depends on the wedge angle, but only becomes reasonably high for very shallow angles, as Table 3.2 shows. This is why most wedges are constructed with relatively shallow angles.

Table 3.2: Mechanical Advantage Generated by a Wedge.

Wedge angle (degrees)	Mechanical advantage
5	11.4
10	5.7
15	3.7
20	2.7
25	2.1
30	1.7
35	1.4
40	1.2
45	1

Chisels

Cutting tools, in the form of stone or flint flakes, are the oldest tools known to man. Most people (including the author of the *Britannica* article mentioned earlier) view chisels as their modern equivalent. This is correct if the chisel is being used to cut wood across the grain. Often, though, chisels act along the grain, and here a chisel is not primarily a cutting tool. It is a wedge. Once the chisel has entered the wood, the shape of the opening crack ensures that the chisel edge is virtually floating free in space, taking little further part in the action (Figure 3.11).

Typical wood chisels have a wedge angle of around 30°. This gives them a mechanical advantage of approximately 2:1 which they don't need after the initial stages, since the process of cutting along the grain is analogous to opening up a crack in any material. The work required to do this depends on the sharpness of the crack tip, not the sharpness of the chisel,

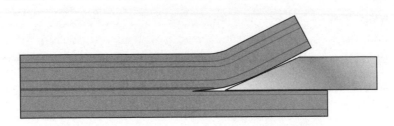

Figure 3.11: Chisel Cutting End-Grain (Showing "Floating" Edge).

except at the very start, when the sharpness of the chisel defines how easy it is to start the crack. After that, the sharper
the crack tip, the less the work, as we saw earlier in the case of
the dry cookie.

If the chisel tip is not sufficiently sharp, it can "catch" in the
opening split and tear at the wood fibers, making the job
much more difficult. So, how sharp does the chisel tip have to
be? That depends on the shape of the bent part of the wood
near the tip — the part that is under the most stress.

To find the optimum shape, I turned to *Formulas for Stress and
Strain*, a book of engineering formulae that I discovered in the
psychology section of a secondhand bookstore. The calculations were complicated, but in the end I found to my surprise
that a chisel edge must be no more than 0.15 micrometers
thick if it is not to catch the adjacent wood. Three hundred
such edges could fit side by side on a human hair. It is no wonder that my father was so angry when I used one of his carefully sharpened chisels as a lever.

When a properly sharpened chisel is used to split wood
along the grain, its main task (once the cut has started) is to
slice any wood fibers that happen to have spanned across the
gap. This action is more efficient if the edge is very sharp and
if the chisel moves across the fiber in a shearing action, which
is why my father taught me to slide the chisel slightly sideways
as it progressed through the wood. That's all that there is to using a chisel to cut along the grain. The trick is to *stop* the sharp
edge from cutting into the adjacent wood, since this will take
the developing split off line. The slightest tilt of the handle

towards the beveled side will result in the edge "catching" the wood. The obvious remedy is always to exert a slight force in the opposite direction, so that the tip remains in free space, out of contact with the wood.

The Screwdriver

Screwdrivers, one of the few hand tools that were not invented until medieval times, are difficult to categorize. If a screwdriver blade is used with a brace and bit, the assembly acts as a lever in the same way that a wrench does. Most screwdrivers, though, are handheld, and their sole purpose, as Jeff Odell pointed out in our coffee-room conversation, is to provide a convenient linkage between the screw and the hand of the user. A screwdriver thus used acts as a rigid extension to the operator's arm, but does not of itself convey a mechanical advantage if properly aligned with the screw. What it does do is to permit the user to attach himself or herself firmly to something that does convey a mechanical advantage — namely, the screw.

A screw driven into the side-grain of a piece of wood acts as a wedge, which is why I have listed the screwdriver in this section. The wedge happens to be wrapped as a spiral around a central shaft, but it is nevertheless a wedge, with the job of levering the wood fibers apart. This is a very difficult task to perform directly (try pulling a piece of wood in half across the grain with your bare hands). Even a softwood such as white spruce has a breaking tension across the grain of 33 kg/cm²; hardwoods can require three or four times as much. The mechanical advantage that a woodscrew provides can be worked out by imagining the thread to be unwrapped and stretched out at the same angle as the original pitch. A "typical" woodscrew from my collection, for example, has a pitch of 5 turns per centimeter and a median diameter of 5 millimeters. From simple geometry, the total thread length (per centimeter length of screw) is 7.8 centimeters, and the mechanical advantage is thus 7.8 (Figure 3.12).

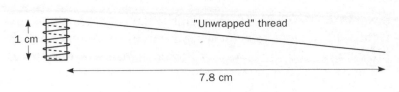

Figure 3.12: "Unwrapped" Screw Thread

The area of wood penetrated by this screw is 0.2 sq. cm, and the force needed to separate the wood fibers across this area is thus $0.2 \times 33 = 6.6$ kg. The mechanical advantage of 7.8 provided by the screw reduces this figure to $6.6 / 7.8 = 0.85$ kg, a figure that would rise to several kilograms if the screw were being driven into hardwood.

I wondered whether I could generate such a force using a twisting motion of my arm only. To find out, I wrapped a rope several times around my wrist, and tied the free end to a series of progressively heavier boulders. I found that I could lift boulders weighing up to 3 kilograms fairly easily with a twist of the wrist (corresponding to a torque with my 7-centimeter diameter wrist of 0.1 kg.m), but that the job thereafter became progressively harder, with 10 kilograms being about my limit. From this experiment, it seems that the mechanical advantage provided by woodscrews is pretty well optimal for the job.

The figures above are approximate, coming from a "back-of-the-envelope" calculation that takes no account of increasing friction between screw and wood; the fact that the central core of the screw is not part of the wedge; or the work needed to displace the wood to make way for this core. The last problem can, of course, be avoided in practice by drilling the wood first to make way for the core of the screw.

Can the job also be made easier by tilting the screwdriver to provide extra leverage? It was time for another experiment. I selected a woodscrew with a slotted head 7 millimeters in diameter, and two screwdrivers of different shaft length, but each with a tip that was a neat fit to the screw slot, and proceeded to use each screwdriver in turn to drive the screw into a piece of softwood. There was no doubt that the longer (25-

centimeter) screwdriver was much easier to use than the shorter (10-centimeter) screwdriver, and that as the screw became more difficult to drive I found myself spontaneously tilting the screwdrivers up to 20° from the vertical, in a direction perpendicular to that of the slot (Figure 3.13). How much extra torque did this tilt provide? A 20° tilt will take the handle of the longer screwdriver some 9 centimeters from the screw axis, but the lateral force that I might have been applying to generate a turning couple is difficult to estimate. Even if that force was only 0.5 kilograms, the torque generated would have been 0.02 kg.m, which adds twenty percent to that provided simply by rotating the wrist — a not-insignificant improvement. A similar calculation shows that the corresponding improvement for the shorter screwdriver is only eight percent, which explains the difference between long- and short-handled screwdrivers.

What, though, is the chance of the screwdriver slipping when used in this way? Not much, as the scale diagram below shows. What about the chance of bending the screwdriver tip? If the lateral force on the handle is half a kilogram, the mechanical advantage is of the order of 250 (!) for a 25-centimeter screwdriver in a 1-millimeter slot, which means

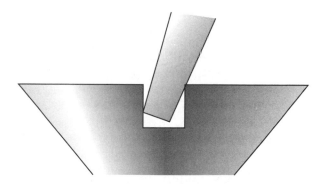

Figure 3.13: Screwdriver Tip in Slotted Screw Head, with Screwdriver Tilted at 20°.
Scale diagram. The screwdriver tip would be a neat fit to the screw slot if the screwdriver were held vertically. The diagram shows that such a tip is unlikely to slip out even if tilted at 20° to the vertical.

72

that the screwdriver is pressing on the side of the slot with a force of 125 kilograms. It sounds like a disastrous scenario (which it might be for the screw head), but the screwdriver is unlikely to be affected, since modern screwdrivers (so Stuart Burgess informs me) are designed to be strong enough to lever the lid off a can of paint. The screw, usually made of a soft metal, is more likely to be damaged.

My argument with Jeff Odell ended up as an honorable draw so far as using a handheld screwdriver is concerned. The best way to drive in a recalcitrant screw, though (apart from using a hammer), is to use a brace and bit fitted with a screwdriver head. The span of a typical brace and bit is around 20 centimeters, which is about the same as that of a large wrench, and provides a mechanical advantage of around 30 for the screw that I have been discussing. The only disadvantage of a brace and bit, so long as it can fit in the working space, is that it makes the job too easy.

Percussive Tools

Hammers

With three of my four initial questions answered, it was time to turn to the fourth — what is the best way to use a hammer? Is there a best weight of hammer for a given nail? How hard should you swing a hammer at each blow? Certainly not as hard as I did when learning as a child. The resultant blow on the thumbnail created an excruciating pressure which my father released by using a needle warmed in a blowtorch flame to red heat to drill a hole in the nail. To this day I am grateful for that rapid piece of amateur doctoring, but it would have been better if I hadn't tried to hit the nail so hard in the first place.

The problem was that I was trying to start the nail off with a single heavy blow. Experience rapidly taught me that a nail needs to be started off with a series of light blows. The reason for this is that, unless the blow is absolutely accurate, there will be a small sideways component to the force generated by the blow, sufficient to knock the nail sideways or for the hammer to glance off the nail head if the blow is too hard.

The idea that a force, or a movement, can be separated into two independent components is one that is not intuitively obvious. I once checked out the question with a group in our village pub, and found that some 30 percent believed that, if I freewheel on a bicycle at constant speed down a hill and throw a stone vertically as I ride, the stone will land behind me (for a full account of this story see chapter six). The correct answer is that it will land beside me, because its forward velocity at the moment of leaving my hand is totally independent of the upward velocity with which I throw it, and so the stone will keep moving forward at the same speed as the bicycle even after it has left my hand.

Scientists are now accustomed to the idea that any movement, or any force, can be regarded as the sum of two other forces at right angles to each other. Luckily, it is easy to work out how large the two different components are with the aid of a simple diagram. The trick is simply to regard the original force or movement as the long side (the hypotenuse) of a right-angled triangle, and just fit the other two sides to it. The result of a nail being struck at a slight angle is shown in Figure 3.14, where the direction of the arrows gives the direction of the forces, and the length of the arrows shows how strong those forces are.

Once the nail has been driven a few millimeters into the wood, the horizontal component of an off-axis swing will be balanced by the restoring elasticity of the wood rather than the grip of the fingers on the nail, and the hammer need not be swung quite so slowly. But how fast should it be swung? Is there an optimum velocity?

When a hammer is swung, some of the energy goes into recoil as the hammer head bounces back off the head of the driven nail. My first thought was that the hammer should therefore be swung slowly for maximum efficiency, with less

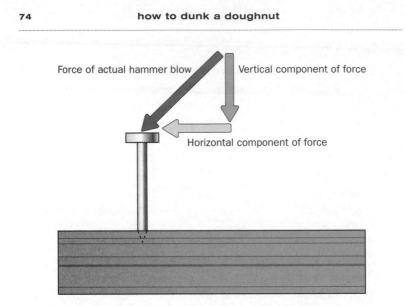

Figure 3.14: Triangle of Forces for a Hammer Swung "Off-Vertical."

recoil and more time in contact with the head of the nail. This time, I didn't need Jeff Odell to tell me that I was wrong. I worked it out for myself, realizing that even if some of the energy that goes into the swing is expended in the recoil, it is still not wasted, since it saves an equivalent amount of the work involved in lifting the hammer for the next blow.

Some hammers are now designed with a layer of polyurethane on the head that "gives" slightly when the nail is struck, keeping the head in contact with the nail for a longer time and allowing more of the energy to be transferred. According to the above argument, this is a gimmick. The only problem in driving a nail in is to maximize the downward component of the force and minimize the sideways component. Even this becomes less of a problem as the nail progressively enters the wood; so the correct technique is to gradually increase the power of the blows, which is something that most carpenters do by instinct. Most carpenters also drill a lead hole, slightly smaller in diameter than that of the nail. This has the advantage of ensuring that the nail is aligned more rapidly by the initial blows, and also that the wood is less likely to split. Another surprising advantage to drilling a lead hole is that the driven nail is harder to remove from the wood than if no hole had been drilled. The reason is that, with no lead hole, the wood is only in actual contact with the nail on two sides. If a lead hole has been drilled, the contact is all the way around and the frictional force is correspondingly greater (Figure 3.15).

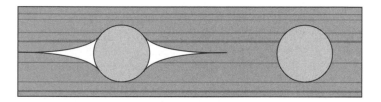

Figure 3.15: Contact of Driven Nail in Wood Without (Left) and With (Right) a Lead Hole.

Only one refinement remains — which part of the hammer head should be used to strike the nail? Most people would say the middle — and most people would be wrong. Hammers actually have a "sweet spot" (very similar to the "sweet spot" in a cricket or baseball bat or a tennis racket), where the jarring is least and the momentum transfer is at its maximum. Its technical name is the "center of percussion," and its existence arises from the fact that the hammer head is swinging in an arc rather than straight up and down. Formulae exist for calculating the center of percussion, which, for something like a bat, can be a lot farther from the hand than the center of gravity. For a hammer, however, my rough calculations showed that the difference was negligible, and that the main problem with using a hammer remains in swinging it accurately in the first place.

*   *   *

At the end of my survey of hand tools, I was disappointed to find that their development, unlike many other commonplace activities, has contributed comparatively little to our understanding of scientific principles. I was very pleased to find, however, that scientific principles are very useful when it comes to using the tools in the most efficient manner.

4
how to add up your
supermarket bill

Physical concepts are not the only barrier between scientists and non-scientists. Numbers and calculations can provide an even greater barrier. Scientists are usually at ease with them, but many other people are not.

The scientist's ease comes partly from familiarity, but also from the shortcuts he or she has learned, which make it relatively easy to juggle with numbers. Scientists apply these shortcuts to scientific calculations. I wondered whether it might be possible to apply them to other areas of life, and decided to try my hand at using them to check bills and assess the pricing policies of my local supermarkets. I was amazed at what the calculations revealed. The tricks are so simple that anyone can use them, both to become more at ease with handling numbers and to check up on what is going on in their own supermarket.

My wife, Wendy, and I live in a small rural English village, but we have several supermarkets nearby, and that is where we do our shopping. Wendy actually does most of the shopping, and when I sat down to write this chapter, I told her that I had devised a statistically based, scientific, simple-to-use method to let her keep a rough track of what she had been spending as she went around buying. She laughed and replied that she already had a method, which was to round down all of the prices ending in ".49" or less, round up the rest, and keep a running total of the results. I told her that it sounded like an interesting approach, but that I was sure my method would be more accurate. You can probably guess the next bit. When I checked the two methods out, Wendy's was a clear winner. It looked as if this was going to be a very short chapter.

Hoping to save something from the wreck, I sat down to

analyze just why her method worked while mine didn't. The first clue came from a friend's supermarket bill, but the final piece of the puzzle was put in place only after I had surveyed nearly a thousand supermarket prices. It transpired that Wendy's approach took almost perfect account of the fact that supermarkets distribute their prices in a very selective way. My supposedly more rigorous statistical approach had failed to cope with this factor, which, as I will show later in this chapter, applies equally to some major American supermarkets.

Armed with this knowledge, I was able to adapt my approach to give the right answer, although Wendy still argues that her method is easier to use. That is up to the reader to decide. More important, I found that the pricing policies used by some supermarkets can be turned, judo-like, against them by the shopper in pursuit of the best value for money.

Taking Care of the Pounds

Supermarket bills in England look pretty much the same as they do in America, except that the English prices are in pounds and pence, rather than dollars and cents. Luckily, the numerical additions are just as easy, since there are one hundred pence in the pound, just as there are one hundred cents in the dollar. So £1.49 + £2.51 = £4.00, just as $1.49 + $2.51 = $4.00. I will talk in dollars and cents where it is possible to do so without ambiguity, but the actual bills that Wendy and I looked at will obviously be in pounds and pence.

The English word *pounds* permitted Lewis Carroll to perpetrate one of the more outrageous puns in the English language. In chapter 9 of Carroll's *Alice in Wonderland*, the Duchess advises Alice to "take care of the sense and the sounds will take care of themselves," which is a wonderful multiple pun on the old English saying, "Take care of the pence and the pounds will take care of themselves." Poorly paid English scientists, though, know that pounds are worth a lot more than pence, and concentrate on the pounds first when it comes to adding up their bills. Wendy and I both used this as our starting point. We ignored the figures that didn't matter and concentrated on those that did.

The figures that matter are called *significant figures*. If a supermarket bill comes to $45.21, for example, the most significant figure is the 4, representing forty dollars of the shopper's hard-earned money. The next most significant figure is the 5 — that extra five dollars matters to most people. The 2 and the 1 hardly matter at all — few people would worry about the extra 21 cents. So far as significant figures are concerned, we can just call the bill $45 and be done with it.

The principle of significant figures is very useful in keeping a running total of what you have spent in a supermarket. Dollars are more significant than cents, and pounds are more important than pence, so a simple way to keep a rough running total is to ignore the cents or the pence entirely. The process is called *truncation* — literally "cutting short." It could also be called guillotining.

The advantage of truncation is that it is an easy feat of mental arithmetic. Its disadvantage is that it only gives what scientists call a first-order approximation to the real total. In other words, it provides a rough first guess with the mental proviso "could do better." In the real bill shown in Figure 4.1, for example (the reader will guess that the flowers were for my wife), truncation gives a total of £5, i.e., 0 + 0 + 2 + 3, compared to the real price of £7.38.

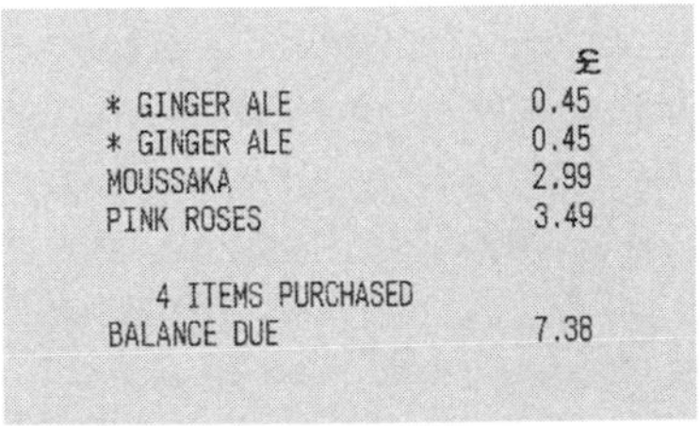

```
                              £
        * GINGER ALE        0.45
        * GINGER ALE        0.45
        MOUSSAKA            2.99
        PINK ROSES          3.49

        4 ITEMS PURCHASED
        BALANCE DUE         7.38
```

Figure 4.1: Short Supermarket Bill.

The basis of my statistical approach is that truncation can lead to better things. That rough first guess is a *lower bound* to the real total, i.e., the real total could not possibly be less. It is also possible, and equally simple, to use truncation to calculate an upper bound to the real total, simply by adding the number of items purchased. In the sample bill, the total obtained by truncation is £5, and there are four items in the bill. The upper bound is therefore £(5 + 4) = £9. This total is the same that would have been obtained by rounding up the price of each item to the next highest pound before adding, because every time you round up the price of an item, all you are doing is adding "one" to the price that you would have gotten by truncation.

Knowing the upper bound to your expected bill provides quite a handy test at the checkout. If the checkout price is higher than the one you worked out in your head, then the checkout total is wrong. An item may have been entered twice, or a price may have been wrongly keyed in. Whatever the cause, it's worth checking.

Upper and lower bounds "box in" the real total between them. Because they are often relatively simple to calculate, scientists frequently use them in the manner of a pincer movement to isolate and trap an otherwise elusive number that would be difficult to pin down in hand-to-hand combat. Archimedes, for example, when not engaged in designing and building war machines, used the "boxing in" technique in his relentless pursuit of the value of π (pronounced "pie"), a number that he needed to know accurately so that he could work out the areas of circular spaces and objects. The ancient Babylonians had known that the area of a circle of radius R is $\pi \times R^2$, and took the value of π to be 3, although they weren't sure whether π was the same for big and small circles. Greek mathematicians prior to Archimedes proved that π had a constant value, but were not much closer to knowing what that value was. Two thousand years later, the Indiana state legislature is said to have come within one vote of resolving the difficulty by declaring the value of π to be 3.2. Unfortunately for ease of

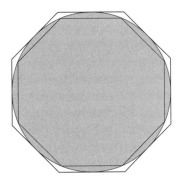

Figure 4.2: Circle Inscribed and Circumscribed by a Pair of Eight-Sided Polygons (Octagons).

calculation, circles are more wont to obey the laws of Nature than the laws of Indiana. Archimedes did much better by drawing two polygons, one circumscribing the circle and one circumscribed by the circle.

It is easy to calculate the area of a polygon, since a polygon can always be divided up into a set of triangles, and the formula for the area of a triangle is well known. Archimedes took advantage of this ease of calculation, together with the facts that the area of the outer polygon is larger than the area of the circle (and hence provides an upper bound to the value of π), while the area of the inner polygon is lower than that of the circle (and hence provides a lower bound to π). By drawing polygons with more and more sides, approximating more and more closely to the shape of the circle, Archimedes approached π from above and below, eventually finding (after drawing a polygon with 96 [!] sides) that π could not be more than 3.1429 or less than 3.1408. The actual value (correct to five significant figures) is 3.1416.

Supermarket Mathematics

Archimedes could have done even better by averaging his upper and lower bounds. This gives an answer of 3.1418 — close enough for most practical purposes. My plan was to use a similar approach to estimate the total of a supermarket bill. In the bill above, for example, the average of the upper and lower bounds (£9 and £5 respectively) is £7, an answer that is satisfactorily close to the true total of £7.38.

A simple, practical way to do this calculation is to keep track of the dollars or pounds (i.e., truncate) while going around a supermarket or adding a bill, and then add half the number of items in the bill to that total. This is mathematically equivalent to averaging the upper and lower bounds (the reason is given in the notes to this chapter), but is much easier to do in practice. It

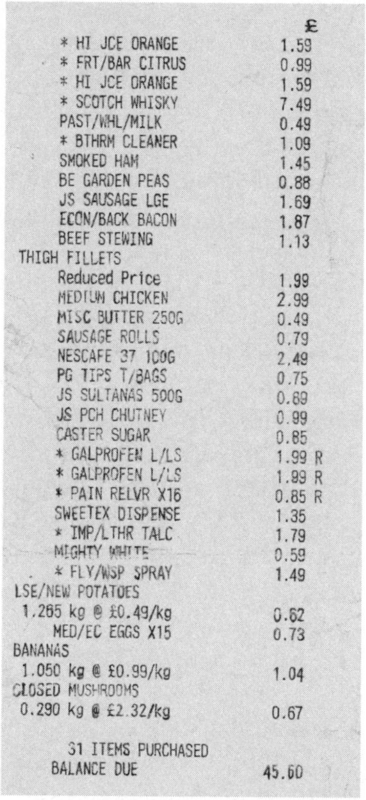

		£
*	HI JCE ORANGE	1.59
*	FRT/BAR CITRUS	0.99
*	HI JCE ORANGE	1.59
*	SCOTCH WHISKY	7.49
	PAST/WHL/MILK	0.49
*	BTHRM CLEANER	1.09
	SMOKED HAM	1.45
	BE GARDEN PEAS	0.88
	JS SAUSAGE LGE	1.69
	ECON/BACK BACON	1.87
	BEEF STEWING	1.13
	THIGH FILLETS	
	Reduced Price	1.99
	MEDIUM CHICKEN	2.99
	MISC BUTTER 250G	0.49
	SAUSAGE ROLLS	0.79
	NESCAFE 37 100G	2.49
	PG TIPS T/BAGS	0.75
	JS SULTANAS 500G	0.89
	JS PCH CHUTNEY	0.99
	CASTER SUGAR	0.85
*	GALPROFEN L/LS	1.99 R
*	GALPROFEN L/LS	1.99 R
*	PAIN RELVR X16	0.85 R
	SWEETEX DISPENSE	1.35
*	IMP/LTHR TALC	1.79
	MIGHTY WHITE	0.59
*	FLY/WSP SPRAY	1.49
	LSE/NEW POTATOES	
	1.265 kg @ £0.49/kg	0.62
	MED/EC EGGS X15	0.73
	BANANAS	
	1.050 kg @ £0.99/kg	1.04
	CLOSED MUSHROOMS	
	0.290 kg @ £2.32/kg	0.67
	31 ITEMS PURCHASED	
	BALANCE DUE	45.60

Figure 4.3: The Supermarket Bill that My Friend Produced.

also seemed to work at least as well as Wendy's method — or so I thought. My confidence in this approach was such that I made a small bet on its accuracy while having a drink with a friend. He immediately produced a crumpled bill from his pocket and challenged me to have a go. I have kept that bill as a reminder that hubris, an occupational disease among scientists, can strike at any time. As Figure 4.3 shows, the bill contained thirty-one items. Half of thirty-one is fifteen and a half, and the total in the "pounds" column is £25, so, according to my rapid approximation technique, the overall total should have been £40.50. Unfortunately, the real total was £45.60. Even more gallingly, Wendy's approach predicted a total of £46.

Supermarket Statistics

Why was my answer so wrong? More to the point, why was Wendy's so right? Both methods, after all, seemed to rely on the same assumption, which was that the average price in the "pence" column over a large number of items would be close to 50p ("p" is short for "pence," just as "¢" is short for "cents"). In my method, this assumption meant that the pence total (expressed in pounds) would be equal to half the number of items. In Wendy's case, the assumption meant that her errors from rounding up would cancel out those from rounding down. In any case, we both seemed to be in the same statistical boat. Things became even more puzzling when I looked at the actual average of the pence column in the bill. This turned out to be 66p. Prices in the pence column were thus clearly biased towards the upper end of the range. It was no wonder that my method had given an answer that was far too low. Why, though, didn't Wendy's approach give an answer that was equally low? It could only be that the errors introduced in rounding up were still being balanced by the errors introduced in rounding down.

When I looked at the bill in detail, I found that this was indeed the case. In her rounding-up range (50p–99p), the average pence price was 81p, so that each rounded-up price would be in error by 19p on average. The average price in her rounding-down range (0p–49p) was 35p, so that each rounding down would, on average, be in error by 35p — almost twice as much as the error introduced by rounding up. This imbalance, which might have been expected to wreck her method, was almost exactly compensated for by the fact that there were twice as many items to be rounded up as there were to be rounded down.

I wondered whether this was a one-time event, or whether the same pattern was followed by other bills from this supermarket. There was only one way to find out, and that was to examine a range of bills, preferably from different shoppers. With the help of my village neighbors, I began to amass a collection. Over the next week piles of bills poured in through the mailbox, to the bemusement of our cat, whose cat door is situated directly below. I also checked the prices for a different supermarket chain in a different town, where unfortunately I had no friendly neighbors to help me. I solved that problem by rooting around in the wastebaskets outside the checkout counters for discarded bills. In the end, I was able to enter nearly a thousand individual prices in my Excel spreadsheet — over six hundred from the first store, and three hundred and fifty from the second, where suspicious looks from the manager eventually forced me to give up my search of his wastebaskets. As a pattern began to emerge, I experienced the same sense of mounting excitement that I have often felt in the scientific laboratory — the special thrill of chasing down and eventually making sense of a set of results.

That feeling, far removed from the often sterile way in which results are eventually presented for public consumption, is the ultimate reward in science. I have seen the most sober-minded of scientists on these occasions sing, dance, and, on one memorable occasion, strip naked and do handstands for the sheer uninhibited joy of the moment. These moments happen because of the tension that builds up during an experiment or series of experiments. That tension wouldn't be there if science operated in the way that many people seem to think it does, with the scientist patiently and objectively collecting "data," and only then thinking about what those data might

mean. But that's not the way most experimental scientists operate. Most of us start off with a picture that we are trying to
check out. The picture may be vague, or it may be very precise. Either way, every fresh result is something to be thought
about and incorporated, either as reinforcement or as a step in
a new direction. As more results come in the excitement
builds, reaching a peak when quick calculations show that the
results are going according to plan. The notebooks of Robert
Millikan, who was awarded the 1923 Nobel Physics Prize for
measuring the value of the charge on the electron, are full of
comments like "Beauty. Publish this surely, beautiful!" and
"This is almost exactly right and the best one I ever had!!!"
My own notebooks are full of similar comments, although I
have taken more care with them since the occasion when my
habitual Australian adjectival expressions were faithfully recorded by my assistant alongside the results.

The picture that I was trying to check out with my supermarket bills was whether the pence prices of the goods in our
local supermarket were really biased towards the upper end of
the scale, and whether there were more actual goods priced
in the upper half of the scale than there were in the lower half.
The act of entering a long string of prices into a spreadsheet
may seem boring to some, but I felt a quickening pulse as I
noted the number of prices that were not only in the upper
half of the scale but actually ended in .99 As I continued to list
the prices from different bills I also noticed that many of the
prices that did not end in .99 ended in .49, and of those that
did neither, many still ended in 9. This was hardly an original
observation, but the scale of this pricing policy was staggering,
with 67 percent of all prices ending in 9.

Then I began to notice gaps. Prices never seemed to end in a
0, 1, or 2, and prices ending in 3, 4, 6, and 7 were noticeably
rare. When I turned to bills from the second supermarket, I
found a similar pattern. To view the data as a whole, and to
look for other patterns and for differences between the supermarkets, I drew a frequency distribution of the prices for each
store (Figure 4.4).

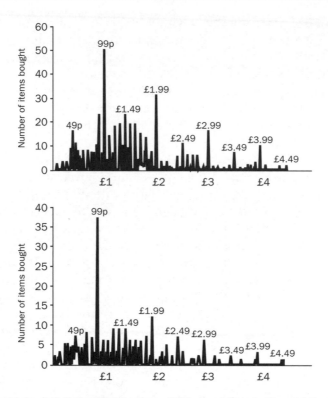

Figure 4.4a and 4.4b: Distribution of Prices of Goods Bought from Two Different Supermarkets.

If the prices in the pence columns had been randomly distributed, then the graphs should not have had any peaks, let alone the huge ones that actually appeared at regular intervals (£0.99, £1.99, £2.99, etc.), decreasing in height with increasing price, and with smaller peaks distributed about them. It appeared that both supermarkets priced many of their goods as close to each whole pound as they could manage without actually touching or going over that barrier. Looking more closely, I observed a second set of peaks, almost invariably higher than their neighbors, but not as high as the first set. These are the peaks at £0.49, £1.49. £2.49, etc. They occur just below the middle of the range (£0.50, £1.50, £2.50, etc.) for any particular pound band. With a high proportion of prices concentrated in these two sets of peaks, it appears that that the fundamental unit of price is actually 50p, and that the price of many goods changes on average by 50p at a time, with the rest of the prices randomly scattered about this mean.

At the higher end of the supermarket price range (not shown in the graphs), this conclusion proved to be exactly right. For items priced above £6, there were no prices that did not end in .49 or .99 in the bills that I examined. Looking at the lower prices, though, it appeared that there was a *third* series of peaks, separated by 10p, with many prices at 29p, 39p, 49p, 59p, etc. This pattern was especially evident for prices between £1 and £2.

It seems fairly obvious, then, and comes as no surprise, that those who set supermarket prices perceive prices ending in zero as barriers not to be passed, or even touched, but to be approached as closely as possible from below. The major barriers are the whole pounds. Less important, but still significant, are prices ending in £0.50. Even prices ending in other multiples of 10p, though, still need to be taken into account for a full picture.

The psychological rationale behind this sort of pricing policy is presumably that we tend to truncate but not to round up when looking at prices. I was more interested, though, to see whether the pence average in the supermarket bill that I had started with was representative of local supermarket bills generally. It was. The average pence price for goods below £2 (representing 80 percent of the purchases of the average shopper) was 60p for the first supermarket and 63p for the second supermarket, with 55 percent of the prices in the upper half of the range for the first supermarket and 64 percent in the upper half of the range for the second supermarket. This suggested that the second supermarket was the more expensive, a conclusion that was later borne out. The figures also showed that Wendy's calculation technique would be expected to work almost perfectly for bills from the second supermarket, and would still be better than mine for bills from the first supermarket.

I was not beaten yet, since I could modify my approach to take account of the new statistical information. The overall pence average for both supermarkets was 61p, which meant that I could get a more accurate total than before by adding up the pounds column and then adding 0.61 times the total number of items instead of half the number. This is not a procedure that is likely to appeal to the average shopper, but a reasonable approximation to it is to add two-thirds of the number of items to the pounds total. In the supermarket bill on page 82, for example, the total number of whole pounds is £25, and there are thirty-one items. Two-thirds of 31 is approximately 20 (20.67, to be more exact), and £25 + £20.67 = £45.67, which is pleasingly, if a little luckily, close to the actual total of £45.60. Usually the overestimate of the total will be slightly greater — but then, it is only an estimate.

Will It Work for American Supermarkets?

My basic ideas for checking and comparing supermarket prices should work anywhere, but the exact numbers to use obviously depend on the pricing strategy of the particular supermarket. I looked up the price lists for ten different American supermarket chains on the Internet, and examined the prices of around a thousand items overall, to see whether these supermarkets adopted similar pricing strategies to British supermarkets. There were some very interesting differences.

The American chains concentrated some of their prices in a similar way to British supermarkets, with major price peaks just below each whole dollar (e.g., 99¢, $1.99, $2.99, etc.). These peaks account for some 30 percent of items in British supermarkets but for around 54 percent in the average American supermarket. In some cases the percentage is even higher. For one large supermarket chain, *every* unit price ended in .99! Clearly my "two-thirds" formula for quickly checking a bill wouldn't work here. On the other hand, this pricing policy lets the buyer estimate the bill simply by rounding up each price to the next dollar and adding. Alternatively, just add the dollars column and then add the number of items.

For other supermarkets, there were two major peaks in the prices — one where the price ended in .99, and one where the price was in whole dollars, i.e., ending in .00. For most of these supermarkets there were nearly twice as many items with prices ending in .99 as there were with prices ending in .00. My "two-thirds" trick for quickly checking a bill should work well in these cases.

Comparing Prices

Consumer organizations, and supermarkets themselves, compare prices using a "standard shopping basket" of the same goods. This can become a game of minimizing the prices in the standard basket, while raising other prices to counterbalance the loss. The individual consumer, armed with the information from my surveys, can do better in a number of ways.

The first is to compare the average prices in the "cents" column for bills from different supermarkets. This comparison technique works best if it is confined to items priced under $5, since the prices of goods that are more costly almost invariably end in ".99." It has the advantage that the shopper can use his or her own "shopping basket" (which need not be identical between the two stores for the procedure to work satisfactorily) to make the comparison. The difference in the "cents" averages might, of course, reflect differences in the quality of the goods. It might also reflect different prices for the same quality — that is up to the shopper to decide. At least, with these calculations, the shopper will be better armed to make the decision.

For a fair comparison, at least fifty items should be included in the total. This usually means saving a number of bills, and the average shopper may not think that it is worth the effort. Is there an easier way? There is, based on the fact that 90 percent of the prices in a typical American supermarket end in ".99" or ".00." This simple pricing strategy means that, if a supermarket is going to raise the price of an item marked in whole dollars, it has virtually no choice, mathematically speaking, but to jump the price all the way up to the next ".99." If the original price ends in ".99," the store has the option of raising the price by just one cent to the next whole dollar, which is a much lower percentage increase in price. So the cheapest supermarket is likely to be the one with the highest proportion of items marked in whole dollars, because the big price jumps for these items haven't happened yet. It's that simple.

Florence Nightingale believed that "to understand God's thoughts, we must study statistics, for these are the measure of His purpose." Be that as it may, there is no better weapon than statistics for understanding and defeating the purposes of supermarket pricing controllers. The quick methods that I have suggested for checking and comparing supermarket prices rely on very simple statistics. I am aware that in suggesting their use in shopping I am in the position of a man who has learned to swim from a book but has not yet tried out his method in practice. There is a simple reason for this — I hate shopping, and will pay anything to get out of the store as quickly as possible. For those with more patience, the simple checks that I have outlined in this chapter should be of some value. Keep in mind that the differences in prices might reflect genuine differences in the quality of the goods, rather than different prices for the same quality. That is up to the shopper to decide. With these simple tests, however, at least the shopper no longer needs to shop in the dark.

Summary

To Quickly Check the Addition of a Supermarket Bill

Add two-thirds of the number of items to the total in the "dollars" column (unless it's one of those supermarkets where everything ends in ".99").

To Compare Prices Between Supermarkets

If most of the prices at the supermarket end in ".00" or ".99," check what proportion actually end in ".00." The higher the proportion, the cheaper the supermarket is likely to be. If the price spread at your local supermarket is more uniform, just look at the prices ending in ".99." The higher the proportion, the more expensive the supermarket is likely to be.

5
how to throw a boomerang

Not many people get to invent a new category for the *Guinness Book of World Records*. I was lucky enough to do so when I was briefly affiliated with a television science program and asked to come up with an idea for a science-based national competition that would attract public attention.

"Why not," I suggested, "get them to design and fly their own boomerangs?" The producer patiently pointed out that outdoor filming was very expensive, and that the idea was to have a competition where the final could be held in a TV studio. "Fine," I said, "we'll get them to make the boomerangs out of cardboard, and have a race to see how many times they can throw them around a pole and catch them in one minute."

The competition was a great success, and a "new" world record was duly established by the British thrower Lawrence West, who zipped his boomerang around a pole three meters away a staggering twenty times in one minute. Our attempts to explain just why boomerangs come back were frustrated, though, when the producer would allow only a few seconds of airtime for the explanation. This was a pity, because the explanation is fascinating in itself, and goes far beyond the confines of spinning boomerangs. In fact, it applies to any spinning object, from the atoms in our brain to the wheels of a bicycle and the Earth in its orbit. All of these respond in the same way when something tries to tilt them over. In the case of a boomerang, the tilting force comes from the air rushing over the "wings." How it happens, and how it makes the boomerang come back, are the subjects of this chapter. As a bonus, there is even a design for a "world record" cardboard boomerang to try.

"Boomerangs? You mean the thinking person's Frisbee?"

The speaker was Sean Slade, secretary of the British Boomerang Society. I had called him in desperation for help with a one-day course on "Boomerangs: Physics and Flying," where I planned to show non-scientists how to design, build, and fly their own boomerangs. I had read up on the physics of boomerangs, but, with only a week to go, I still hadn't managed to fly one. No matter how hard I tried, my boomerang wouldn't come back. Sean was surprised, to put it mildly, to encounter an Australian who didn't know how to throw a boomerang, and offered to introduce me to the art. It is an art that Australian Aborigines have understood for at least ten thousand years, although modern science is only just starting to come to grips with it.

The next day, Sean turned up on my doorstep with a bag containing more than 150 boomerangs. Some were made of wood, the traditional material, but many more were made from the materials of modern science — brightly colored plastics, carbon fiber–reinforced composites, and even metals like titanium. The longest was almost as tall as him, and looked a fearsome weapon.

Returning boomerangs were, in fact, hardly ever used as weapons. Australian Aborigines, like modern-day boomerang throwers, used (and still use) them for sport. The sport could be quite bloodthirsty, as the following account from 1881 shows:

> Ten or twelve warriors, painted with white stripes across the cheek and nose, and armed with shields and boomerangs, are met by an equal number at a distance of about twenty paces. Each individual has a right to throw his boomerang at anyone on the other side, and steps out of rank into the intervening space to do so. The opposite party take their turn, and so on alternately, until someone is hit, or all are satisfied . . . As the boomerang is thrown with great force, it requires very great dexterity and quick sight to avoid such an erratic weapon, and affords a fine opportunity for displaying the remarkable activity of the aborigines. This activity is, no doubt, considerably roused by fear of the severe cut which is inflicted by the boomerang.

Some modern boomerangs can be equally frightening. Sean told me that a new world distance record had just been established — not by an Aboriginal thrower, but by a Frenchman competing in a contest at Shrewsbury in England. The record distance was an incredible 149 meters, but the triumph was nearly a catastrophe because a good distance boomerang is like a flying razor blade, being very thin and sharp-edged to reduce aerodynamic drag and to permit it to travel as far as possible before returning. It is practically invisible to the thrower as it returns, edge-on, and this one nearly removed the thrower's head, eventually landing some sixty meters behind him.

Despite their potential for damage, returning boomerangs, as mentioned, are seldom used as hunting weapons because it is much easier to hit a target such as an animal or an enemy with a direct throw of a spear or a stone, rather than with something that not only follows a widely curving path, but which also rises as it flies. When Australian Aborigines use boomerang-shaped sticks as weapons for war or hunting, the sticks (called *kylies* in some Aboriginal languages) are deliberately chosen *not* to return. The arms are shaped to have aerodynamic "lift" so that when the stick is thrown horizontally, spinning at around ten revolutions per second, it skims a meter or so above the ground in a straight path until it hits whatever is unfortunate enough to be in the way. Some 95 percent of the boomerang-shaped objects that have been collected fall into this category.

Boomerangs themselves are different. Their arms are also shaped to give aerodynamic "lift," but the lift operates in such a way that when the boomerang is thrown in an initially near-vertical orientation, it gradually turns over as it travels in a circle until, by the time it returns to the thrower, it is spinning in a horizontal plane and hovering, just waiting to be caught.

As I looked at Sean's large boomerang, and even those of more modest proportions, I thought with some trepidation that they would be pretty lethal even when hovering, and could easily take a couple of fingers off. He demonstrated the way to avoid this, which was to catch the boomerang between the palms of the hands by clapping them together. This turned out to be easier than it looked, and was the start of my love affair with boomerangs.

Records are plentiful in boomerang throwing, which is now an organized sport, with competitions for "fast catch," "distance thrown" (this was also a competition among the Aborigines), "maximum time aloft," and numerous other categories. The grand prize for "maximum time aloft" must surely go to Bob Reid, a British university physicist who managed to keep a boomerang in the air for twenty-four hours and eleven seconds. Impossible? Not if you are enthusiastic, and eccentric, enough to take a boomerang to the South Pole and throw it through all the time zones, thus adding twenty-four hours to the actual time of flight (Figure 5.1).

Figure 5.1: Dr. Bob Reid in the Process of Throwing a Boomerang Around the South Pole.

With people such as Sean and Bob for support, I felt that I was in my element as I settled down to address the question "What makes a boomerang come back?" The short answer is spin, together with the fact that the two blades of a standard boomerang are oriented differently (Figure 5.2).

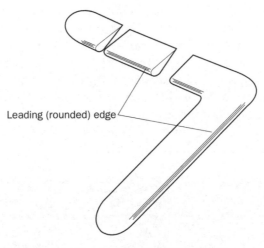

Leading (rounded) edge

Figure 5.2: Wing Shapes in a Traditional Boomerang.

Each blade is shaped like an airplane wing, but the two blades are joined together in such a way that the rounded edge of both wings always leads as the boomerang spins after it is thrown. Both wings thus provide "lift" but, because the boomerang is spinning in a vertical plane, the "lift" pushes the boomerang sideways, ultimately returning it to the thrower.

I hope that the above explanation does not satisfy you. It is the one that I put together for the TV producer in response to nagging demands for a ten-second "sound bite" on how boomerangs work. Scientists are frequently asked for such sound bites, the presumption being that the audience could not possibly remain interested for any longer, or understand anything more complicated. I don't believe it, and I think that audiences are being patronized by such practices.

The real reasons for the return of the boomerang are easier

to understand than the simplified explanation, which raises more questions than it answers, and are also a lot more interesting and aesthetically pleasing. In science, as in art, the beauty is in the detail. Not that visual beauty is missing. Boomerangs themselves have a pleasing symmetry, and are usually painted, although the paintings on pre-European Aboriginal boomerangs were seldom intended for decorative purposes. They were used in the same way as other Aboriginal art is used — as a reference to ancestors and to the Dreamtime when the Earth was formed (Figure 5.3). Most surviving examples have long since been snapped up by collectors, and modern Aboriginal decoration is exclusively for the tourist market.

I came across one curious example when I was purchasing boomerangs from Duncan MacLennan, an old-time Australian who has run a boomerang school in Sydney for nearly sixty years. When he started, he wanted Aboriginal motifs for

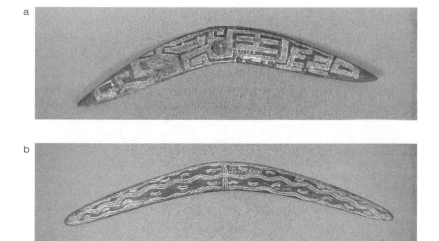

Figure 5.3: Australian Aboriginal Decorated Boomerangs.
a. Kimberley Region of Western Australia.
b. Diamantina River Region. Picture provided by Philip Jones, Senior Curator, South Australian Museum.

the plywood boomerangs that he was manufacturing, and approached the members of a local tribe to do the painting. They were very interested, not only in doing the painting, but also in the boomerangs themselves. It turned out that they had never seen one, and Duncan had the unique pleasure of teaching an Aboriginal tribe how to throw boomerangs.

The science of what happens when a boomerang leaves the hand of the thrower is as aesthetically pleasing as the boomerang itself. The simple key notion is that the two arms, even though they are identical and symmetrically disposed about the axis of spin, experience different aerodynamic forces because the boomerang is moving forward as well as spinning. The net result is that the upper arm is moving faster through the air than the lower arm, so that the sideways "lift" is greater on the upper arm than it is on the lower arm, and the boomerang tilts (Figure 5.4).

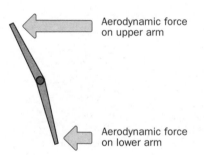

Aerodynamic force
on upper arm

Aerodynamic force
on lower arm

Figure 5.4: Flying Boomerang Viewed Along Flight Path (Central Circle) from Behind.
(The magnitude of the aerodynamic force on each boomerang arm is proportional to the length of the arrow.)

What happens next is that the boomerang's flight direction changes to follow the direction of the tilt. This phenomenon, called *precession,* is one that we often experience, even if we can't put a name to it. When we lean sideways while riding a bicycle, for example, the spinning front wheel automatically turns to follow the direction in which we lean (the back wheel would do likewise if it could). The phenomenon also works in reverse. If we turn the front wheel of a bicycle, we automatically lean in the direction of the turn. This is not voluntary — it's physics.

Precession can turn up in unexpected places. Hospital MRI (magnetic resonance imaging) scanners use it, for example, in imaging soft tissues such as the brain. The things that are spinning are not the patients but the atoms in the patients' brains. An MRI scanner subjects these atoms to a weak radio signal that tilts them over. The spinning atoms respond just as a tilted bicycle wheel would, by trying to recover the direction of their spins. The speed at which they can do this depends on how "sticky" the local brain environment is, which can be different in sickness and in health, so that the speed at which the spinning atoms recover after being subjected to a radio pulse can give valuable clues about whether the part of the brain that they inhabit is healthy or ailing. Precession in boomerangs, bicycles, and brains works in the same way in each case, and is ultimately describable by a very simple and beautiful rule, called the *gyro rule,* which says that *the spin axis chases the torque axis.* It is easiest to explain this rule with a diagram, as in Figure 5.5.

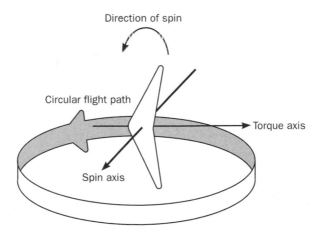

Figure 5.5: The Gyro Rule Applied to Boomerangs.

The spin axis is the line about which the spin is taking place (e.g., the axle of a bicycle wheel). Similarly, the torque axis is the one about which the tilt is taking place. For the boomerang shown in the diagram, the torque axis is the line of flight, with its direction specified by the *right-hand screw rule*. Bend the fingers of your right hand in the direction of rotation or torque. The thumb will now be pointing in the direction of the appropriate axis. The right-hand screw rule also applies to the spin of the boomerang itself. The directions of the spin and torque axes, together with the gyro rule, tell us that boomerangs swing to the left if they tilt to the left, eventually swinging right around and coming back towards the thrower. It is also possible to design a boomerang that tilts to the right when thrown, and therefore turns around in the opposite direction. The first type is usually used by right-handed throwers, and the second type by left-handed throwers. So there are right-handed and left-handed boomerangs, just as there are right-handed and left-handed corkscrews.

The gyro rule follows from one of the great conservation laws of classical physics. There are just four of these, all enunciated in the nineteenth century, and all describing a fundamental quantity whose total throughout the universe remains forever unchanged. One of those quantities is energy (see chapter two). The other three are electrical charge, linear momentum, and angular momentum.[1]

Precession explains very elegantly *why* boomerangs come back, but that is only a start, as Figure 5.6 reveals. The path shown is the simplest, and typical of many boomerangs. Viewed from above, it is a circle. Viewed from the side (as in the photograph), the boomerang rises ever more rapidly as it leaves

---

[1] Linear momentum is just mass × linear velocity (i.e., speed in a given direction). Angular momentum is mass × angular velocity (i.e., rotational speed). Linear momentum is conserved unless a force is applied, and the direction of the force determines the direction of the new momentum. Angular momentum is conserved unless a tilting force (a torque) is applied, and the direction of the torque determines the new direction of the angular momentum. The spin axis defines the direction of angular momentum. When the torque axis is at right angles to the spin axis, then the latter is forever chasing the former.

Figure 5.6: Time-Lapse Photograph of a Boomerang in Flight.
The boomerang, thrown by Michael Hanson and photographed by Christian
Taylor, was lit by a small battery-powered bulb at the tip.

the thrower, then slowly settles as it returns, looping and spinning as it does so. This photograph is the latest (and best) in a
long line that stretches back to a time-lapse photograph taken
in front of the old Australian Parliament House by Lorin
Hawes, an eccentric American physicist who gave up a job as
a nuclear weapons instructor and moved to Australia, where
he set up as a boomerang designer and founder of the wonderful Mudgeeraba Creek Emu Racing and Boomerang Throwing Association.

It seems impossible that such a flight path could be accounted for by simple physical rules. People before Copernicus
had a similar problem with the planets. Viewed from Earth,
they can appear to speed up, slow down, and even occasionally do backwards somersaults. If we could put a large enough
sparkler on Mars and take a time-lapse photograph of its
movements, they would appear remarkably similar to those of
a boomerang. The Greek astronomer Ptolemy explained such
complicated planetary movements in terms of small-scale circular movements (called epicycles) superimposed on the main

orbit, and even smaller-scale circular movements superimposed on these. It took the genius of Copernicus to realize that all of Ptolemy's problems stemmed from taking the Earth as the point of observation. Viewed from the Sun instead, planetary movements became extraordinarily simple, with all planets following smooth elliptical orbits.

Boomerang movements become equally simple when viewed from the point of view of the boomerang. The first step is to explain why the boomerang flies in a circle. A circular orbit requires a force that is always directed towards the center and which does not change in magnitude. This force is provided by the aerodynamic lift. For it to stay constant, the angle of attack of the airfoil (i.e., the angle of the boomerang wing to the onrushing air) must be constant. This will only happen if the rate at which the boomerang tilts is exactly the same as the rate at which it flies around the circle. It takes only three lines of (admittedly complex) algebra to show that a simple, ideal boomerang with wings of a uniform cross-sectional shape obeys this condition. The mathematics, summarized in two excellent articles by the irrepressible Bob Reid, leads to a "boomerang equation" that tells us a lot about boomerangs and how they work. Here I have written the boomerang equation using words instead of algebraic symbols:

$$\text{Radius of flight circle} = \frac{\text{density of material} \times \text{cross-sectional area of boomerang arm}}{\text{lift coefficient} \times \text{width of boomerang arm} \times \text{density of air}}$$

In other words, the flight circle will be bigger if the material from which the boomerang is constructed is denser and if the arms are thicker. The flight circle will also be bigger if the lift coefficient (i.e., the aerodynamic effectiveness of the wings) is smaller, if the arms are narrower, and if the density of the air is lower. So if you want distance, go to a high place, as the Swiss thrower Manuel Schütz did in 1999 when he established a new world distance record of 238 meters (more than a quarter of a mile, there and back!). The record was established at Kloten, near Zurich, 409 meters above sea level, and with an air density four percent less than that at sea level.

Even more interesting than the factors that are in the equation are the factors that aren't. The radius of the flight path is independent of both the forward velocity and the angular spin velocity. It is also independent of the length of the arms, except insofar as this length affects the lift coefficient, and is even independent of the number of arms.

Some of the factors in the boomerang equation fight with each other. A low lift coefficient, for example, usually needs a thin, flat arm, rather than a thick one. Practical boomerang design, then, is more than a matter of fitting an equation, even though the equation provides a useful guide. Compromise, based on experience, can considerably improve the performance.

Whatever the design of the boomerang, the radius of the flight circle is built in, although factors such as the angle of tilt and speed of rotation at the throw can sometimes have a substantial effect. The main result of throwing a boomerang harder, though, is to make it travel around its flight path faster, and come back sooner.

Lawrence West, the current world record holder for indoor boomerang flying, took advantage of this fact in the competition that I organized for the British TV show *Tomorrow's World*. The competition was for the maximum number of times that a cardboard boomerang could be thrown and caught in one minute after passing around a pole three meters away. Lawrence designed a boomerang where the balance of factors in the boomerang equation provided a built-in flying radius of just over 1.5 meters, whipping it around the pole twenty-four times in practice, and twenty times in the actual competition (Figure 5.7).

The science involved some interesting subtleties that I introduced to the participants in my course when they were designing their own boomerangs. The first concerned the stability of the boomerang, and the angle between the arms. It might seem from the boomerang equation that the angle between the arms doesn't matter, since it isn't mentioned. It

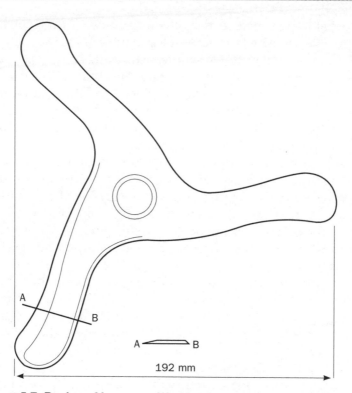

Figure 5.7: Design of Lawrence West's "World Record" Cardboard Boomerang, Made from 1.5 mm-thick Mount Board.

does matter, though, for a reason that is easy to understand intuitively although rather complex to explain scientifically. If you imagine making a straight boomerang, with the two arms in the same line, it is easy to see that this would be very difficult to throw so that it kept spinning in the same plane. It is all too likely also to start rotating about the longitudinal axis, eventually resulting in a crazy tumbling.

Stability and rotation are favorite subjects of my Bristol University colleague Professor Mike Berry, who has established a worldwide reputation for demolishing long-held beliefs in this area. His latest triumph earned him the spoof IgNobel Prize from Harvard University because of its eye-catching description — "Of Flying Frogs and Levitrons." The science, though, was very real. The frog was a toy that seemed to disobey a fundamental physical law by remaining suspended in space. The problem was not the defiance of gravity — this was achieved by putting a set of magnets in the frog, with their north poles pointing downwards, and placing the frog over a base containing a set of magnets with their north poles pointing upwards. The real problem was that such an arrangement has been known for at least one hundred years to be unstable. The slightest touch, or puff of breeze, and the frog should flip over and fall to the ground. Yet here was the frog, displayed in a shop window, lazily spinning for hours on end.

The spin was the key, as Mike discovered when he sat down to do the mathematics of the problem. It transpired that the arrangement is unstable *unless* the frog is spinning, in which case there is a very narrow range of parameters within which the arrangement remains stable. The inventor of the levitating frog, not knowing that it was impossible to levitate a frog by means of magnets, had stumbled on the precise set of conditions needed.

Boomerang designers have a similar problem. The physicist's "ideal" boomerang, with wings of uniform cross-section, makes the job of calculation easier, but isn't particularly ideal when it comes to competition throwing. With any other design, though, the thrower faces the immediate problem of making the rate of tilt equal to the rate of rotation, which is mathematically equivalent to saying that the angle of attack has to stay constant. This means that the rate of tilt (i.e., the precession rate) must be just right. Too slow, and the angle of attack gradually decreases so that, after a promising start, the boomerang won't come back. Too rapid, and the angle of attack increases, the lift increases, and eventually the boomerang stalls.

What can be done with a boomerang that won't come back, either disappearing into the distance or stalling and crashing? Some of the students in my course thought that the answer would be to change the shape of the boomerang by sanding it. The real answer is much simpler than that — just tape small coins to the wings. For a boomerang that stalls, the mathematics underlying the boomerang equation show that the coins should be taped near the tip, slowing the boomerang down. To speed up the boomerang, the coins should be taped near the center. The effect in the latter case is the same as when a spinning skater with arms outstretched pulls his or her arms in, speeding up as a result because angular momentum is therefore conserved. Boomerang designers regularly use the coin trick, only changing the basic design once the coins have shown where weight needs to be added and where it needs to be subtracted.

The other main design problem with boomerangs is constructing them so that they "lie down" at just the right stage in the flying circle. But why does a boomerang "lie down" at all? The answer concerns yet another subtlety of the boomerang, which is that one of the arms is always flying in "dirty air," a phenomenon that will be familiar to yachtsmen, and indeed to anyone who has been passed by a fast-moving truck on a highway. This phenomenon happens because the two arms of a traditionally shaped boomerang, apparently symmetrically disposed, are in fact quite different. The arm with the sharp edge on the "inside" is called the leading arm. The other (with the sharp edge on the "outside") was christened by Lorin Hawes the dingle, or dangling, arm. The poor dingle arm is forever flying through the disturbed air created by the leading arm. This means that if the two arms are identical, the dingle arm will provide less lift than the leading arm. "Averaged over a complete rotation," says Bob Reid, "there is a net torque about the vertical axis, which in turn leads to a precession about the direction of flight, i.e., the boomerang lies down."

After half an hour with Bob's diagrams, I believed him. Others will have to have faith, as my students did when they made their own boomerangs, or read the original article themselves. The main practical point is that the boomerang's propensity to lie down can be controlled by shaping the dingle arm so as to give it slightly more lift than the leading arm. This is easily done with a touch of sanding, which fine-tunes the boomerang so that it will lie down at just the right point, as

Lawrence West's did, enabling him to catch and throw it with machine-like precision.

Most of this I learned when I was planning my one-day course, but all of the theoretical knowledge in the world wasn't getting my boomerang to come back, and, as the time drew near, I was becoming increasingly frustrated. Then I found out what the problem was: air resistance.

A boomerang transfers some of its momentum to the surrounding air molecules as it travels, speeding them up (so the air becomes hotter) but slowing down itself. The drag can be sufficient to slow both the forward motion and the spin to the point where the boomerang simply gives up and flops to the ground. The answer is to get the boomerang spinning as fast as possible at the start, so that it still retains a reasonable proportion of its angular momentum by the time it returns. The secret, as Sean showed me just in time, is to tilt the boomerang right back and then to release it with a whipping action, launching it at a few degrees above the horizontal to counteract the effects of gravity on the flight.

When the time came for the course itself, I was throwing like the expert I wasn't, and the students were following my advice to produce some wonderful boomerangs. I forbore to tell them that Australian Aborigines mostly use their boomerangs as knives, digging tools, musical instruments, and for cleaning their teeth. It would have been a pity to spoil their pleasure as, one and all, their boomerangs came flying back.

6
catch as catch can

Until recently, catching a ball was one of the few areas of sports that science had not touched. Rackets, bats, and the other tools of sports have long been designed along scientific principles, as have athletes' diets and, to some extent, athletes themselves. Many sporting techniques have also been subject to scientific refinement — javelins, for example, are now launched at a precisely calculated angle, with the thrower moving in a scientifically guided way. Ball catching, though, has remained the province of natural skill. The closest that science has come to it is recounted in A. G. Macdonell's classic 1933 account of an English village cricket match, which remains the funniest account of any sporting moment that I have read. The game (in baseball terms) is all tied up before the final out, and the "bowler" is equivalent to a pitcher, while the "wicket-keeper" is the equivalent of a catcher. Now read and enjoy!

The scores were level and there was one wicket to fall. The last man in was the blacksmith. . . . He took guard and looked round savagely. He was clearly still in a great rage.

The first ball he received he lashed at wildly and hit straight up in the air to an enormous height. It went up and up and up, until it became difficult to focus it properly against the deep, cloudless blue of the sky, and it carried with it the hopes and fears of an English village. Up and up it went and then it seemed to hang motionless in the air, poised like a hawk, fighting, as it were, a heroic but forlorn battle against the chief invention of Sir Isaac Newton, and then it began its slow descent.

In the meanwhile things were happening below. . . . The titanic Boone had not moved because he was more or less in the right place, but then Boone was not likely to bring off the catch, especially after the episode of the last ball. Major Hawker, shouting "Mine, mine!" in a magnificently self-confident voice, was coming up from the bowler's end like a battle-cruiser. Mr. Harcourt, the poet, had obviously lost sight of the ball altogether, if indeed he had ever seen it, for he was running round and round Boone and giggling foolishly. Livingstone and Southcott, the two cracks, were approaching competently, their eyes fixed on the ball. . . . In the meantime, the professor of ballistics had made a lightning calculation of angles, velocities, density of the air, barometer-readings and temperature, and had arrived at the conclusion that the critical point, the spot which ought to be marked in the photographs with an X, was one yard to the northeast of Boone, and he proceeded to take up his station there, colliding on the way with Donald and knocking him over. A moment later Bobby Southcott came racing up and tripped over the recumbent Donald and was shot head first into the Abraham-like bosom of Boone. Boone stepped back a yard under the impact and came down with his spiked boot, surmounted by a good eighteen stone of flesh and blood, upon the professor's toe. Almost simultaneously the portly wicket-keeper, whose movements were a positive triumph of the spirit over the body, bumped the professor from behind . . . and all the time the visiting American journalist Mr. Shakespeare Pollock hovered alertly upon the outskirts . . . screaming American university cries in a piercingly high tenor voice.

At last the ball came down. . . it was a striking testimony to the mathematical and ballistical skill of the professor that the ball landed with a sharp report upon the top of his head. Thence it leapt up into the air a foot or so, cannoned on to Boone's head, and then trickled slowly down the colossal expanse of the wicket-keeper's back, bouncing slightly as it reached the massive lower portions. It was only a foot from the ground when Mr. Shakespeare Pollock sprang into the vortex with a last ear-splitting howl of victory and grabbed it off the seat of the wicket-keeper's trousers. The match was a tie.

I can vouch for the essential truth of the above description, since I still play cricket for the English village where I live.

Surprisingly, science now seems to have shown that the professor's ability to rapidly calculate trajectories is something that we all have, and which we use when we run to catch a ball. Support for this idea came in a paper published in 1993 in the prestigious journal *Nature*, where two psychologists analyzed video-records of expert catchers running to catch a ball. The authors found that the catchers varied their running speeds so that the rate at which they were tilting their heads to follow the flight of the ball conformed to a specific equation. They believed that the catchers could only achieve this feat by solving the equation in their heads as they ran, and concluded that this "demonstrates the power of the brain's unconscious problem-solving abilities."

I was initially convinced by their argument, as were the newspapers of the day, one of which even ran the story on its front page. When I looked at what the equation really meant, though, I found that it conveys an absurdly simple physical message about what we have to do to catch a ball. That message is one that we learn as children, and what the psychologists' work really tells us provides a fascinating insight into how we adapt the simple techniques that we learn as children to the more complex situations of later life.

Children begin learning to catch by standing with their arms outstretched, waiting for an adoring adult to lob a ball into their hands. The success of this enterprise is not infrequently spoiled by the fact that most young children tend to shut their eyes and turn their heads away as soon as the ball is thrown. With time, children learn that this is not a particularly efficient technique, and begin (after some encouragement) to watch the ball, eventually learning to move their heads so as to follow its flight.

With repeated practice, children learn to use the rate of head-tilting as a cue to tell them whether they are standing in the right place to make the catch. If they are, then the cue is a very simple one, and can be described by an equation that was worked out by two groups of physicists some thirty years ago.

The authors of the *Nature* paper showed that we continue to use this same cue as adults, even when we have to run to make a catch. How do we use it, though? Do we solve the equation in our heads, or does the equation merely describe some simple physical action that we learn by repetition, in the same way that a musical score might tell us to move our fingers in a certain way on a keyboard, a way that we learn with practice so that we do not therefore need the music to guide us?

To find the answer, I decided to repeat the earlier calculation for myself. I had to use differential calculus, a method that scientists use for tracking rates of change. Its principles are simple — far simpler than those involved in reading music. They were splendidly revealed to me as a child in a wonderful little book (still available) written by the impressively named Silvanus P. Thompson, a Cambridge professor of engineering who had as his motto: "What one fool can understand, another can." That was good enough for my father, who owned the book, and it was good enough for me.

This fool soon understood that the calculus tracks change by dividing the change into small steps. When a ball is thrown, for example, its flight path can be worked out by dividing the horizontal movement into small steps, and working out the effect of gravity on the height after each step. The principle is illustrated in Figure 6.1, where the smooth trajectory of a thrown ball is approximated by eight discrete steps. In this diagram the thrower has launched the ball at an angle of 45° to the horizontal. The ball's upward speed gradually decreases under the retarding influence of gravity, and the upward movement eventually stops. The ball does not disobey "the chief invention of Isaac Newton" because the direction of its motion is immediately reversed, and the ball starts to fall with ever-increasing speed under the influence of gravity. These changes in vertical speed have no effect on the horizontal speed of the ball. It can be quite difficult to see this, as I found when I was called upon to settle an argument on the subject in my village pub.

The discussion that night had touched on the weather, politics,

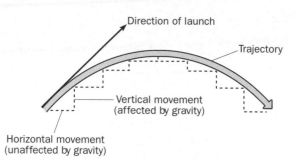

Figure 6.1: Trajectory of a Ball Launched at an Angle of 45°, Approximated by Eight Equal Horizontal Steps.
The vertical movement after each horizontal step is calculated from Newton's Law of Gravity, which says that the force of gravity changes the vertical speed by 9.8 meters per second every second (i.e., the acceleration, or rate of change of downward speed, is 9.8 m/s²).

and the state of the thatched bus shelter in the village square, but had eventually settled in some inscrutable way on the question of what happens if someone riding a bicycle throws a ball straight up in the air. Opinions were divided. Some claimed that when the ball came down it would land beside the rider. The majority view, though, was that the ball would land some way behind. Those who supported the latter view were quite disappointed when I sided with the minority, giving as my reason that the ball would keep moving forward at the same speed as the cyclist. Since several pints of beer were riding on the outcome, I was challenged to fetch a bicycle and prove my point experimentally.

I proved it only too well when I freewheeled down the hill past the cheering crowd at the pub door and launched a small stone that I had picked up from the road vertically into the air. The stone kept pace with me and, a few meters further on, landed directly on the top of my head. I was glad that it hadn't been a cricket ball or a baseball.

An even more dramatic illustration of the independence of horizontal and vertical motions is given by Lewis Wolpert in his book *The Unnatural Nature of Science*. Suppose a marksman fires a rifle bullet horizontally, and simultaneously drops a second bullet from the hand that is supporting the rifle. Both bullets will hit the ground at the same time. The horizontal speed of the first bullet makes no difference at all to its vertical movement.

The independence of horizontal and vertical speeds has been proved repeatedly under controlled laboratory conditions, but this is only half the story when it comes to understanding why the stone landed on my head. The other half concerns why the stone kept moving forward at the same speed once I had let it go. It's not obvious that it should. Common sense suggests that things only move if they are pushed or pulled — in other words, if a force is applied to them. This was the view taken by Aristotle two and a half thousand years ago, and it is the view taken by many people today. According to a recent survey, some 30 percent of people still share Aristotle's commonsense notion that things stop moving once there is nothing to push or pull on them.

It took two thousand years, and the genius of Newton, to discover that things actually stay still *or keep moving at constant speed in a straight line* unless a force is applied to them (Newton's First Law of Motion). My stone was therefore obeying Newton's First Law when it kept moving forward at the same speed after it had left my hand. If I had looked up at it, it would not have appeared to me to be moving in a horizontal direction, since we were both traveling with the same horizontal speed.

From the point of view of the observers at the pub door, though, its movement would have appeared to be very different. They also would have observed the stone traveling forward at a constant speed (i.e., traveling twice as far in two seconds as it had in one second). In the vertical direction, though, the stone was being accelerated downward under the force of gravity. The law of acceleration, discovered by Galileo, says that if the time of travel is doubled, the stone will travel four times as far. Combination of this vertical accelerated motion with uniform horizontal motion produces a parabolic path when viewed from the side.

A catcher standing in the right position to make a catch has a different point of view again. He or she is not viewing the flight path from the side, but sees it "end-on," and has to use the angle of gaze as the main cue to judge the position of the ball. How does this angle of gaze change with time as the ball approaches? We can make an intelligent guess by drawing a diagram of the position of the ball in its parabolic flight at equal time intervals, corresponding to equal horizontal steps in the diagram because the ball is traveling at a constant horizontal speed (Figure 6.2).

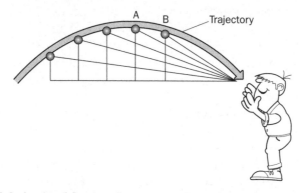

Figure 6.2: Angle of Gaze at Successive Equal Time Intervals for a Catcher Watching a Ball Approach Along a Parabolic Path.

It appears from this diagram that the changes in angle over successive time intervals are pretty much the same up to an angle of 30° or so (line from point A to catcher). In other words, a catcher standing in the right position tilts his or her head at a constant rate to follow the flight of the ball, so long as the trajectory doesn't go too high. Conversely, tilting the head at a constant rate provides a cue which tells the catcher that he or she is standing in the right position. It seems that this is the cue that children learn, and which they continue to use into later life.

It is not a cue that is always reliable. When the angle becomes too high (e.g., line from point B to catcher) its rate of change slows down. If the catcher continues to use the same cue, he or she will interpret this slowing down to indicate that the ball is approaching more slowly than it is, and will miss the catch. This may be the reason why high catches are so often missed.

The diagram above provides a good guide to how the angle of a catcher's gaze changes when the catcher is in the right position, and may be as far as some people would want to go. To me, the full and satisfying answers would only come when I had worked out an equation to describe how the angle changes with time, and when I could understand from that equation whether catchers might need to adapt their technique for different situations. I set to work with a light heart one Saturday morning before a village cricket match, stimulated by a glass of Australian chardonnay and by the thought that I could check out the answers against rough experimental observations in the afternoon. As piles of paper grew around me, each sheet containing one or more mistakes in the horribly complicated algebra, I began to wonder what I had let myself in for. Nevertheless, I pressed on, and was rewarded when most of the terms in half a page of algebra canceled each other out, leaving a beautifully simple solution.

My feeling on finding an elegant mathematical solution was akin to that of an Australian traveler discovering an unexpected country pub. The fact that someone else had found the same solution (albeit expressed in a different form) some thirty years earlier did not detract from my feeling, which was a mixture of triumph and relief — relief at having gotten the answer, triumph that it told me something new about the science of catching a ball.

Those readers who prefer to avoid mathematical symbols entirely can safely skip the next three paragraphs. The symbols tell the story so beautifully, though, that I include them here for those who would like to see what I saw, and in the way in which I first saw it.

When I began the calculations, I expressed the angle of gaze in terms of its gradient, called a "tangent" in mathematical language, and abbreviated to "tan." The term "gradient" has exactly the same meaning that it has on a road sign — in other words, the vertical distance climbed or descended divided by the horizontal distance necessary to achieve the climb or descent. In Figure 6.3, for example, the gradient of the hill is 1 in 4, expressed by saying that tanα = ¼.

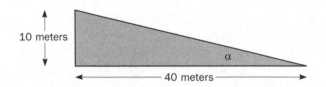

Figure 6.3: A Hill with a Gradient of 1 in 4.

I wanted to find out how tanα changed with time — in this case, the rate at which a catcher's angle of gaze changes. Newton expressed rates of change by putting a dot over the top — I did the same, so that what I was after was $\dot{\tan}α$.

What might $\dot{\tan}α$ depend on? I put in everything that seemed to be relevant, including the angle of launch, the time for the ball to travel from thrower to hitter, and the distance of the catcher from the ball from the thrower or hitter. In the end, all of these canceled out. All that I was left with was the acceleration due to gravity (g) and the horizontal speed of the ball (v). The gloriously simple answer was:

$$\dot{\tan}α = g/2v$$

What does this equation tell us? For a start, it tells us that if a catcher is standing in the right place to make a catch, the gradient of the angle of gaze will change at a constant rate as he or she follows the flight of the ball. Conversely, a constant rate of change in the gradient is a cue that tells us we are standing in the right place.

For angles below about 30°, the gradient of an angle is proportional to the angle, so the angle of gaze itself changes at a constant rate, just as the earlier diagram suggested. For higher angles, the gradient changes more rapidly than the angle, so we have to tilt our heads at a progressively slower pace to keep the gradient changing at a constant rate. If we keep tilting them at the same rate, obeying the lessons of our childhood, we will miss the catch.

The equation also predicts the actual rate at which a catcher's head needs to tilt, which depends on the horizontal speed of the ball. If a ball is approaching the catcher with a horizontal speed of fourteen meters per second (50 km/hr), for example, the catcher needs to tilt his or her head at around 17° per second to follow its trajectory, no matter what the angle of launch. If the horizontal speed doubles, the rate at which the catcher has to tilt his or her head *halves*, making it easier to judge where to place the hands for the catch, even if the catcher's reaction time isn't fast enough to achieve this in practice.

Finally, the equation says that the head has to keep tilting in the same direction until the catch is made. If the direction reverses at any stage, that is a sure clue that the catch is going to be missed, unless it is made below eye level.

The real lessons of the equation, though, come when we consider how we use it as a cue in the process of making a running catch. The authors of the *Nature* paper showed that good catchers control their running speed in this situation so that the simple cue provided by the equation continues to be obeyed. What running strategy, though, must a catcher adopt to do this?

I tried to work it out mathematically, and the result was depressing indeed — a horrendously complicated expression for the way in which a catcher's running speed changes with time. One thing was clear, though — a catcher cannot run at a constant speed and still keep tilting his or her head at a constant rate to follow the ball. To keep the head tilting at a constant rate, the catcher's running speed must change with time in a complicated way that depends on the angle of launch, the speed of the ball, and the distance of the runner from the correct spot. In all cases, the runner is accelerating or decelerating as he or she makes the catch.

When I looked at the physics of the situation, it turned out that this acceleration or deceleration gives the running catcher a clear advantage over someone who is already on the right spot and does not have to move. Running is one of the keys to catching a ball, because a person who is running finds it easier to make small adjustments of position than one who is standing still. The difference lies in the balance of forces that the people experience.

A person who is standing still experiences two major forces: the force of gravity, acting downwards, and the force of the ground's reaction on the feet, acting upwards. So long as these forces stay in line, everything is fine. The balance is precarious, though, because human beings are relatively long and thin, with a high center of gravity, so that balancing a human body is rather like trying to balance a pencil on its end. A tilt of no more than 6° is sufficient to transform the gravitational and reactional forces into a couple that tilts the person further until, like an overloaded wheelbarrow, he or she falls over, unless the person is quick enough to move the feet wider apart or move the arms, like a tightrope walker, to shift the center of gravity.

Some animals, such as tortoises, lizards, frogs, and toads, overcome the balance problem by having widely spaced feet and a low center of gravity. Such animals are not notable ball-catchers, and one may wonder why humans, with their more unstable configuration, are so much more successful at this and other tasks that require coordinated movement.

The answer is that our more unstable configuration makes us much more maneuverable, since a relatively small force can shift our position rapidly and substantially. That force comes from pushing down and back with our feet when we run. This extra force changes the balance situation so that we can tilt but still remain stable, because the net thrust is a diagonal one, so the reaction force passes through our center of gravity even though we are tilted forward (Figure 6.4).

The extra force only cuts in when we are accelerating, which scientists define as changing speed and/or changing direction.

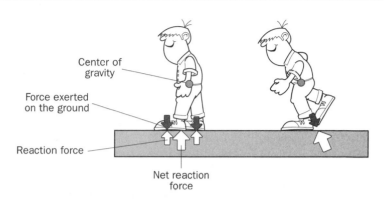

Center of gravity

Force exerted on the ground

Reaction force

Net reaction force

Figure 6.4: Forces When Standing or Moving at a Constant Speed (Left) and When Accelerating (Right).
The black arrows represent the force of our bodies on the ground. The white arrows are the reaction force (equal and opposite) of the ground on our feet. Balance is maintained if the net reaction force passes through our center of gravity.

Running at a constant speed requires surprisingly little force — just enough to overcome wind resistance, friction in the joints, and wasted energy in the muscles. Without the extra force required for acceleration, runners have to adopt a virtually upright position for balance. The exception occurs at the start of a race, when the athlete is accelerating (i.e., changing speed) and can hence lean forward and use the additional force to maintain balance.

According to Newton's First Law, a person who is standing still or running at a constant velocity does not generate the extra force required for balance and maneuverability produced by someone who is accelerating. From this point of view, the best way to catch a ball is not to run at a constant speed to intercept it as it falls, but to be accelerating or decelerating when the catch is made. This is exactly what the "ball-catching" equation predicts, when modified for the case of a running catcher, and is also what the authors of the *Nature* paper found from their experiments.

It is also what I found from watching my fellow village cricketers, and what I observed from watching cricket and baseball on television. There was one exception — professional baseball players appear to run to the correct position and then wait when catching a high fly ball. The practical reason for this is that the final vertical speed of the ball is much greater for a high catch, so that the catcher has to be close to the optimum spot to have any chance of making the catch at all. There is also a more subtle reason, which is that the tangent of an angle begins to increase much more rapidly for angles over 30° (as in baseball), so that someone tilting their head at a constant rate at the start of a ball's flight will be left far behind towards the end of the flight. By getting close to the optimum spot, there is less room for error, which more than compensates in this case for the loss of maneuverability that comes from standing still.

My conclusion, then, is that the running catch is the better technique for balls launched at angles below 45°, while the "get there and wait" technique is better for high fly balls. In both cases, the successful catcher is the one who judges the catch (unconsciously) through the cue of tilting the head at a constant rate, or better still at a rate that keeps the gradient of the viewing angle changing at a constant rate. Rapid mental calculation has little to do with it — we actually use very simple cues indeed as we keep our eye on the ball.

7
bath foam, beer foam, and the meaning of life

The story of science is not simply a matter of high points and great peaks of discovery. The scientists who reached those peaks generally did so by dogged perseverance and by focusing on the detail each step of the way. Stories of the pursuit of this detail can be just as fascinating as stories of the great achievements themselves, although they undoubtedly make more demands on the reader because of the amount of background explanation necessary to understand what that detail means. The following chapter tells one such story. It is the story of a relatively minor advance in science, concerning how molecules "self-assemble" to form complicated structures such as foams. The results now permeate every corner of our lives, from the way we wash our hair to the way we administer drugs. Our understanding of the processes involved has even influenced our view of how life itself was originated and evolved on Earth.

The story is told from my perspective as a privileged inside observer. By talking about it in more detail than is usual in a popular science book, I hope to convey something of what it feels like to be a real scientist doing real science, where the beauty resides much more often in the day-to-day detail than in the grand conceptions that the public hears much more about.

According to the American scientist Sidney Perkowitz, the twenty-first century will be remembered as the "Foam Age." Aluminum foams will be used to make cars with light, strong bodies. Concrete foams that support normal loads but crumple under heavy weights have already been made, and will form the ends of airport runways to safely slow down airplanes that have overshot on takeoff or landing. NASA, Perkowitz said, has even launched a rocket with an ultra-lightweight foam section designed to capture particles from the tail of a comet.

Perkowitz's foams have bubbles with solid walls that give them their strength. The foams that I discuss in this chapter, such as beer foam and bath foam, have liquid walls. How those walls form, and how they maintain their strength, has been a perennial puzzle that goes right back to Newton and his experiments with soap bubbles. I like to think that he experimented with them in his bath, but the truth is that he probably blew them in a glass bowl of soapy water placed on a table. What he saw was what we have all seen but, being Newton, he was able to take his observations of a common phenomenon just that one step further, as he later reported in his 1704 book on *Opticks:*

> If a Bubble be blown with Water first made tenacious by dissolving a little Soap in it, it will appear tinged with a great variety of Colours. To defend these Bubbles from being agitated by the external Air (whereby their Colours are irregularly moved one among another, so that no accurate Observation can be made of them,) as soon as I had blown any of them I cover'd it with a clear Glass, and by that means its Colours emerged in a very regular order, like so many concentrick Rings encompassing the top of the Bubble. And as the Bubble grew thinner by the continual subsiding of the Water, these Rings dilated slowly and overspread the whole Bubble, descending in order to the bottom of it, where they vanish'd successively.

Newton knew that the colors depended on the thickness of the soap film, which varied from top to bottom. He also noted that "Sometimes the Bubble would become of a uniform thickness all over . . ." By comparing the bubble color at this stage with the colors produced by gaps of known thickness between two pieces of glass, he was able to calculate the thickness of a soap film that had drained to a uniform silvery color as "3⅛ ten hundred thousandths" of an inch — in modern units, 80 nanometers (80 billionths of a meter), which means that his soap film was around five hundred water molecules thick. That's about ten times thinner than can be observed with the naked eye. How does such a gossamer-thin film remain so stable? Newton thought that it was because the soap dissolved in the water and made it "tenacious." It was another two hundred years before it was recognized that soap molecules prefer to inhabit the surface of water, and exert their magical effects mainly at that surface.

The person who first recognized why detergent molecules behave in this way was Irving Langmuir, a scientist at the General Electric Company in Schenectady, New York, who had the enviable brief of studying anything that he might find interesting. At various stages in his career he discovered a way to greatly extend the life of the tungsten filament lamp, developed new vacuum tubes for use in radio broadcasting, and was awarded a Nobel Prize for his ideas on how atoms bind together to make molecules. During the Second World War he became involved in rain-making experiments that had the unexpected consequence of one American state threatening to sue another for the theft of its rain. Underpinning it all was Langmuir's interest in how molecules behave at surfaces (one modern journal devoted to this topic is now called *Langmuir*). He began by studying molecules on solid surfaces, but eventually turned his attention to liquid surfaces, where, in the early 1930s, he quickly realized that detergent-type materials prefer to reside at the surface of water because their molecules have chemically different ends. One end, called the *hydrophilic* (water-loving) end, is chemically similar to water, and has a preference for being in water. The other end, called the *lipophilic* (oil-loving) end, is chemically similar to oil, and prefers an oily environment. If there is no oil around, air will do almost as well. Scientists who have become bored with long Greek words simply call the hydrophilic and lipophilic ends *heads* and *tails* respectively.

The peculiar structure of detergent molecules makes them *surface active* — if shaken up in water, they will move to the surface, where they sit like feeding ducks with their hydrophilic heads in the water and their oily tails sticking up in the air (Figure 7.1). As many detergent molecules as possible will seek to occupy the surface, jostling for position and eventually becoming packed together as closely as they can manage. Those left under the water (the majority) will still seek to arrange themselves so their tails are hidden from the water while their heads are immersed in it.

Figure 7.1: Monolayer of Detergent Molecules (Represented as "Ducks" with Hydrophilic Heads and Hydrophobic Tails) on the Surface of Water.

There are two basic ways that they can do this. The first is to arrange themselves in a ball (called a *micelle*), with the tails in the middle and the heads at the surface (Figure 7.2). Micelles can take grease into their oily centers, and do most of the work in washing up. The other type of structure that detergent molecules can arrange themselves into underwater consists of a pair of flat sheets (called a *bilayer*), aligned so the heads point

Figure 7.2: Detergent Molecules Packed in a Ball (a "Micelle") Underwater.

outwards with tails on the inside, again hidden from the water (Figure 7.3).

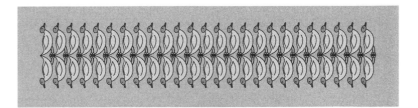

Figure 7.3: "Bilayer" of Detergent Molecules Underwater.

The problem with such a bilayer is that the molecules at the edges still have their tails exposed. This problem is overcome when the bilayer spontaneously curves around on itself to form a closed sphere called a *liposome* (Figure 7.4). These, like

Figure 7.4: "Liposome" of Detergent Molecules Underwater.
Liposomes are often encapsulated by concentric shells of bilayers. Technically, a liposome consisting of just one bilayer is called a *vesicle*, but I have avoided using the term in this chapter in favor of the more familiar generic term *liposome* to cover all types.

micelles, can carry materials in their centers, hidden from the water outside. The difference is that micelles carry oil-soluble materials such as grease, whereas liposomes can carry materials that are soluble in water.

The membranes that encapsulated the first living cells were probably liposomes formed spontaneously from lecithins manufactured naturally under the conditions that then prevailed on Earth. There is great interest in using such liposomes to carry drugs to their site of action in the body while keeping them hidden from destructive enzymes en route. The targeting is achieved by incorporating marker molecules into the outer wall of the liposome. The challenge is to do this in such a way that the liposome bilayer structure is not disrupted. This means understanding the forces that hold the liposome together, and how these forces might be affected by the introduction of foreign molecules into the bilayer structure. It also means understanding how the shapes of detergent molecules affect the way they pack together to form different structures. These two lines of thought were brought together in a grand synthesis in the late 1970s. Up until then, though, they had developed as almost separate lines of inquiry.

How Do Detergent Molecules Stick Together?

The first line of inquiry concerned the forces between detergent molecules, and how these forces convey stability — not just to soap films, but to a much wider class of materials called *colloids.*

I was introduced to colloids in the early 1960s by Professor Alexander E. Alexander (invariably known as Alex), my favorite university lecturer and a thorn in the side of the Australian education authorities. Scarcely a week passed without a strongly worded letter from Alex appearing in at least one Sydney newspaper, usually about the lack of state support for scientific education or the iniquitous behavior of some member of the education bureaucracy. Alex's fellow professors frowned on his practice of opening up the closed academic world of the time to public scrutiny, but Alex couldn't have cared less. He was having too much fun in his own world, populated by scientists who were almost as gregarious as he was. When I eventually met and worked with some of them, I already knew quite a bit about their personal foibles through the anecdotes that peppered Alex's lectures.

Alex was the coauthor of an authoritative textbook on colloids, and won the hearts of his undergraduate audience by declaring that this massive tome was a book that "need never have been written." Nevertheless, it was loaded with interesting, if not always relevant, information, such as the fact that *kolloid* was the ancient Greek word for glue. In modern terminology, a colloid is simply a suspension of small particles in a medium. Milk is a colloid, because it is a suspension of milk fat globules in water, and so is paint, a suspension of solid pigment granules in oil or water. Cigarette smoke is a colloid, because it is a suspension of ash particles in air (smoke will appear white if the particles are large enough to reflect light from their surfaces, and bluish if the particles are so small that all they can do is scatter the light). Blood, a suspension of living cells in serum, is also a colloid, together with a host of other materials vital to life.

The defining feature of colloids is that the particles are small and consequently the total surface area is huge. The surface area of the milk fat globules in a pint of homogenized milk, for example, is around two hundred square meters — about the floor area of an average house. Forces between the surfaces of the particles thus dominate the behavior of a colloidal suspension. These forces (called *surface forces*) arise largely from the single layer of molecules at the surface of each particle. Normal red cells, for example, repel and slide past their neighbors because their surfaces are coated with a layer of negatively charged sugar molecules. Milk fat globules stay apart for a similar reason, except that in this case the protective molecules are natural milk proteins.

In the absence of the repulsive forces provided by protective layers, similar particles will stick together because they are pulled towards each other by a universal attractive force of electrical origin called the *Van der Waals force,* which increases rapidly in strength as the particles get closer to one another. This can cause problems. If milk fat globules stick together, for example, the milk separates into a layer of cream above and a layer of water below. If red blood cells stick together, they can no longer squeeze through tiny capillaries, and the capillaries become blocked. This happens to people suffering from sickle-cell anemia, one of the "molecular" diseases in which the alteration of one small group of atoms in the hemoglobin molecule causes that molecule to adopt a different shape, leading in turn to the deformation of the whole red cell. The molecular changes also make the outside of the cell slightly sticky, so that instead of repelling and sliding past its neighbors, it adheres to them, especially in tight corners such as those in joints, where the cells clump together, blocking blood flow and causing excruciating pain.

Particles such as milk fat globules or red blood cells will only stay apart if the repulsive force is sufficient to overcome the attractive Van der Waals force at some stage in the approach. If the repulsive force is not present naturally, it can be added. One of the principal ways of doing this for suspensions of hydrophobic particles in water (e.g., grime in bathwater) is to add detergent molecules. The hydrophobic tails of the detergent molecules anchor themselves to the particle surfaces, thus hiding both from the surrounding water, while the electrically charged heads of the detergent molecules form a protective outer layer that repels other, similarly coated particles and keeps them in a loose suspension that can easily be rinsed away instead of collecting as a sticky layer on the sides of the bath.

The idea that a balance of attractive and repulsive forces controls colloid stability was developed independently in the early 1940s by two groups of scientists, Boris Deryagin and Lev Landau in Russia and Evert Verwey and Theo Overbeek in Holland. Both groups published their ideas after the Second World War and, after a brief battle over priority, the theory became democratically known as the Deryagin-Landau-Verwey-Overbeek theory, universally abbreviated to "DLVO."

Deryagin dominated Russian surface science for some seventy years, and made many important discoveries sometimes wrongly attributed to Western scientists, owing to the slowness with which Russian publications were disseminated and accepted in the West. He also made more controversial claims. When I heard him give his last major conference talk in Moscow in 1992, he laid claim to the debatable phenomenon of "cold fusion," which he had attempted to initiate in a typically ingenious manner by having his students fire a Kalashnikov rifle at the experimental apparatus from close quarters. Deryagin seldom left Russia, but he did make one visit to Australia when he was in his eighties, although he was still active — active enough, in fact, to be dropped off in the red light district of Sydney at ten o'clock at night, not to reappear until three the following morning, by which time his hosts were beside themselves with anxiety.

The Dutch scientists Verwey and Overbeek were much more sedate when I first met them in 1976, and very generous with their time and expertise to a junior, rather brash researcher from Australia. Theo Overbeek even offered to give his lectures in English rather than Dutch for my benefit. I was impressed, but declined. I would have been even more impressed if I had known that he spoke five other languages as well.

DLVO theory was just that — a theory. It needed testing, and soap films seemed to be an ideal vehicle, since the repulsive forces due to the charged head-groups limited how thin the water could get. The Dutch group in particular performed many experiments with soap films, varying the ultimate film thickness by adding different amounts of salt to the water. DLVO theory predicted that the salt would reduce the repulsive force between the charged head-groups on opposite sides of the soap film, and so allow the film to become thinner. Measurements of the film thickness using reflected light produced numerical values that were very close to those predicted (Figure 7.5).

Studies of soap films eventually yielded a great deal of information about the repulsive forces between detergent head-

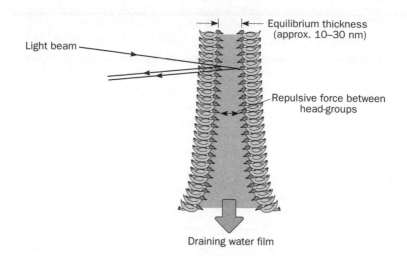

Figure 7.5: Molecular Detail of Draining Soap Film Being Examined by Reflected Light.

groups, but did not contribute directly to solving the problem of why some detergent molecules spontaneously form flat films while others form structures like micelles. Nevertheless, when the answer did come, it turned out that the repulsive forces measured by the soap film enthusiasts were a key component.

One of the problems with soap films is that the force tending to thin the film comes from hydrostatic pressure rather than from Van der Waals forces, which become important only when the gap between the two surfaces is less than ten nanometers or so (a gap that could be spanned by a chain of about sixty water molecules). Soap films are usually considerably thicker than this, and new techniques had to be developed to measure forces at smaller distances.

All of these techniques made use of ultra-smooth solid surfaces mounted on springs stiff enough to keep them apart under the large close-range attractive forces predicted. The Russian school measured the forces between crossed wires, while the Dutch school used polished glass lenses. The advantage of glass was that the experimenter could see through it, and could use reflected light to measure the distance between the surfaces. Its disadvantage was that glass surfaces are relatively bumpy at an atomic level. Wire is smoother, but it is impossible to measure the separation distance directly. The stage was set for a fiery confrontation, which duly occurred at a conference that I unfortunately missed, although the reverberations were still being felt when I entered the field some ten years later.

By that time a new technique had been developed by David Tabor in Cambridge, England, who used mica, which was transparent and which could also be cleaved to atomic smoothness. A succession of Ph.D students refined the technique, with the final, crucial step being taken by Jacob Israelachvili, who performed the astonishing feat of controlling and measuring the distance between the mica surfaces to within one atom. He later told me with some relish that when he had reported his results at a scientific meeting in America one of the older-style scientists in the audience had sat stolidly through his presentation, and then told him that it was physically impossible to do such measurements.

Jacob's original measurements were made in air. I met him when he had moved to Australia and had begun to refine his technique to make similar measurements with water between the surfaces, which he was eventually to coat with detergent and other molecules. Our meeting was pure serendipity. I happened to drop in on a talk that Jacob was giving about his new technique, and rapidly realized that it would be ideal for the totally different problem on which I was then working. I outlined my idea over coffee after the talk. The result was a collaboration that lasted several years and totally changed the direction of my research.

Every few months over those years I would drive or fly the three hundred kilometers from Sydney to Jacob's laboratory in Canberra to indulge in an intense week of experimentation, often working through until two or three in the morning. Situated in an old wooden cottage on the shores of Lake Burley Griffin, the Department of Applied Mathematics where Jacob then worked was rapidly becoming one of the great world centers of surface science. Coffee-room conversations seemed to cover almost every aspect of the field, questioning old shibboleths and producing new ideas in profusion.

The department was founded and run by Barry Ninham, a mathematician of extraordinary versatility who had chosen surface science as his field, and who had already written a book on the theory of Van der Waals forces, so complete that the theory need never be touched again. Now Barry was looking for new fields to conquer. Jacob provided an ideal opportunity when he began to talk about his ideas of how detergent-type molecules pack together in different ways to form micelles, liposomes, and other structures. The secret, he believed, lay not just in the forces between the molecules, but also in the shape of the molecules within these structures. Those that packed in layers, he thought, would need to have heads with approximately the same cross-sectional area as the tails so that they could fit easily into a planar structure. Those that packed more easily into micelles, though, would have larger heads, and so would pack more naturally into spheres, with the heads on the ouside where there is more room (Figure 7.6).

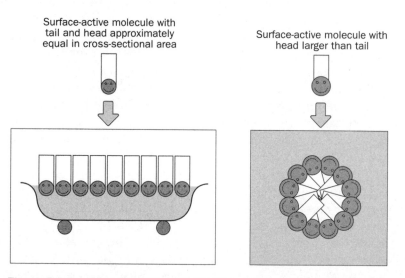

Figure 7.6: Packing Strategies for Surface-Active Molecules with Different Shapes.

How Do Molecules Shape Up?

The idea that molecules have three-dimensional shapes goes back to Louis Pasteur, whose deduction in 1844 arose from the fact that solutions of two chemical compounds with apparently identical composition could nevertheless twist a beam of light in opposite directions. The difference, he thought, must be because the molecules of the compounds have different three-dimensional shapes.

Pasteur was right, and his observations led to the modern subject of *stereochemistry,* the science of molecular shapes and how these shapes affect molecular behavior. It was the subject of my very first research project, and nearly my last. The project was to measure the shapes of some small molecules called sulfoxides dissolved in benzene, a highly inflammable solvent. The technique, called the Kerr effect, followed Pasteur in using the twisting of a light beam to provide information about the molecules that could be used to deduce details of their shapes. To make the light rotate sufficiently, though, the molecules had to be aligned in a strong electric field, which means putting ten thousand volts across the solution.

In such a situation, a spark would have caused a catastrophic explosion just a couple of centimeters away from my eye, and the benzene had to be thoroughly dried to prevent this from happening. I did this by the approved technique of adding fresh sodium metal to take up the residual water. Unfortunately, I failed to notice that a speck of sodium was still present when I disposed of some residual benzene down the sink (a procedure that would never be allowed these days). The sodium reacted vigorously with the water in the pipes, producing a jet of flaming hydrogen that in turn set fire to the benzene, sending a scorching flame up the laboratory wall and nearly putting paid to my scientific career before it had fairly started.

Indirect approaches like the Kerr effect have been largely replaced by direct techniques, such as *X-ray crystallography,* that permit the experimenter to measure the positions in space of all of the individual atoms in a molecule. X-ray crystallography,

which works with any material that can be persuaded to form
a solid crystal, however tiny, was the technique that permitted
the helical structure of DNA to be worked out in the early
1950s. The technique has now progressed so far that scientists
can even use it to watch enzymes swallow their molecular
prey and regurgitate it in a different form in real time.

Even more exciting than X-ray crystallography is the new
technique of *scanning probe microscopy,* which permits scientists
to see individual molecules — or, at least, to feel them. The
technique is similar to that used by a blind person who waves a
white cane back and forth on the pavement as he or she walks.
The cane senses bumps and dips in the path, and could in principle be used to map its contours. Scanning probe microscopy
does the same thing at an atomic level, with the "cane" being a
miniature pointed tip attached to a tiny spring, whose deflec-
tions map the surface to atomic resolution. The modern scanning probe microscope is a lovely little instrument, about the
size of an electric toaster, and can be used to view just about any
molecule, from relatively small detergent molecules to huge biological molecules such as DNA (Figure 7.7).

Figure 7.7: Scanning Probe Microscope Picture of DNA Molecule.

Early workers studying the shape and size of detergent molecules did not have the advantage of such techniques, and had to develop other approaches. These usually involved spreading the molecules as a single layer on the surface of water. The very first measurement of the size of a molecule worked in this way and was developed fortuitously by the American scientist and statesman Benjamin Franklin.

The story began on a sea journey in 1757, when Franklin noticed that the wakes behind two of the accompanying ships were smooth, while those behind his own and the rest of the ships were rough. He called the attention of the ship's captain to this remarkable phenomenon, and later reported the captain's contemptuous reply in a letter to a friend: "The Cooks, says he, have I suppose, been just emptying their greasy Water thro' the Scuppers, which has greased the Sides of those Ships a Little."

Franklin thought that it was much more likely that the oil was having an effect on the waves rather than the ships, but wisely kept his thoughts to himself. He tested this idea on and off over the next seventeen years, culminating with the experiment he performed on a pond in London's Clapham Common. The wind was ruffling the water when he tried pouring a teaspoon of olive oil onto the surface. He later wrote to his friend William Brownrigg:

the Oil tho' not more than a Teaspoonful produced an instant calm, over a Space of several yards square, which spread amazingly, and extended itself gradually, until it reached the Lee Side, making all of that Quarter of the Pond, perhaps half an Acre, as Smooth as a Looking Glass.

Franklin thought that he had discovered a method for calming rough seas, and kept proposing the method thereafter, even though his later large-scale experiments at Spithead in northwest England produced nothing more than the world's first oil slick. What he had in fact done was develop the first method for measuring the size of a molecule.

Franklin didn't realize what he had really done — he didn't even know that molecules existed. Nevertheless, his experiment keeps turning up in modern-day examination papers, usually with the volume of the teaspoon chosen so as to give the right answer. The plain fact is that we don't know exactly how big Franklin's teaspoon was. What we do know is that a film of oil such as that which he described keeps on spreading until it is just one molecule thick, but does not spread further because the molecules in the film are held together by Van der Waals forces. If the volume and the area of the film are known, the thickness can be calculated. A hundred and sixteen years later, Lord Rayleigh repeated Franklin's experiment "in a sponge bath of extra size," using a carefully measured volume of oil and floating pieces of camphor to mark the boundary of the film. He concluded that the thickness of the film was 1.6 "micro-millimeters" — in modern units, 1.6 nanometers. This means that Franklin's teaspoon must have had a volume of about 3 ml, which is approximately that of a Georgian teaspoon in my possession.

The key step in turning Franklin's and Rayleigh's experiments into a proper, scientific tool for studying the shape and size of surface-active molecules was taken in 1891, the very next year, by one of the few women scientists of the time, the German Miss Agnes Pockels, who discovered that a surface film of such molecules could be compressed by a sliding barrier. It was not until 1935, though, that a working scientific instrument based on this principle was built by the American scientist Irving Langmuir. This was developed in collaboration with another woman scientist, Dr. Katherine Blodgett, but neither Blodgett's nor Pockels' contribution is acknowledged in the modern name for the instrument, which is simply called a Langmuir trough. The nomenclature is not Langmuir's fault — he was one of the few scientists of his time who apparently showed no regard for status or gender, but treated everyone on an equal footing. The Langmuir trough (not even mentioned among Langmuir's scientific achievements in his entry in the *Encyclopaedia Britannica*) is simply a flat shallow trough filled to the brim with water, and with a dilute layer of the surface-active molecules spread on the surface. The layer is compressed by a sliding barrier, with the experimenter using a spring-mounted hanging plate to monitor the response of the layer to compression (Figure 7.8).

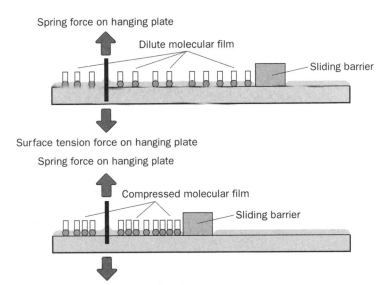

Spring force on hanging plate

Dilute molecular film

Sliding barrier

Surface tension force on hanging plate

Spring force on hanging plate

Compressed molecular film

Sliding barrier

Surface tension force on hanging plate

Figure 7.9: Schematic Side View of Langmuir Trough Before and After Compression of Molecular Surface Layer.

The Langmuir trough provided two vital pieces of information. The first came from the change in area as the barrier was moved, which told the experimenter how closely the molecules were packed. Eventually the packing density reached a limit that depended on the shape and size of the individual molecules. A "typical" detergent molecule has a cross-sectional area of about 0.25 square nanometers.

The second piece of information came from the change in surface tension (registered by how far the spring-mounted hanging plate was pulled down) as the film was compressed, which gave a measure of the work needed to push the surface molecules together, and hence gave information about the forces between those molecules. For charged surface-active molecules, the net force is repulsive. Given enough room, the molecules will get as far away from each other as possible, like relatives at a wedding.

I was introduced to the Langmuir trough technique by Alex, who was an authority on its use, and who taught us that the trough and the water in it must be scrupulously clean. Modern troughs are made from Teflon, the polymer material that is used to coat nonstick pans. Teflon is relatively easy to keep clean, but earlier metal or plastic troughs had to be coated with a protective material as a barrier against contamination. Alex and his Cambridge-trained contemporaries used cheap paraffin wax for this purpose, but some other experimenters really went to extremes. Alex was particularly scathing about an American colleague named James William McBain, who had coated his entire apparatus with gold.

The trough sounds very simple, but was quite tricky to use, because the amount of material required to cover the surface completely is invisible to the naked eye — a visible speck would have been far too much. The trick lay not in wearing the strong magnifying spectacles that go with the traditional image of a scientist, but in dissolving the material in a relatively large volume of a volatile solvent, such as ether, and then putting a small measured drop of the solution on the trough's surface. The ether evaporated, leaving the material behind as a single layer of molecules (a monolayer) covering the surface of the trough.

The Langmuir trough had by now been used to settle many outstanding questions about the shapes and packing of molecules at surfaces. One of the earliest of these concerned the shape of the cholesterol molecule, which is one of the basic molecules of life. Cholesterol receives bad press these days because of its role in forming plaques that can block blood vessels and cause heart attacks or strokes, but it causes harmful effects only when present in excess. The effects would be even more harmful if you didn't have enough of it — you would be dead. The shape of the cholesterol molecule dictates how it behaves chemically in our bodies. At the time when the Langmuir trough was invented, the atomic composition of the cholesterol molecule was known, but the atoms could have been arranged in two quite different ways, with each structure corresponding to a different molecular shape. Each shape had its proponents, who were happily arguing in the knowledge that there was no experimental method available to settle the argument. The two proposed structures predicted quite different cross-sectional areas for the cholesterol molecule, and the famous surface scientist N. K. Adam realized that the difference would be shown up clearly by the newly invented Langmuir technique, because one structure was flat, so that the molecules would stack like plates arranged vertically, side by side on the water surface, while the other structure predicted a more "three-dimensional" shape, so that the area per molecule would be much greater. It took just one experiment to resolve the argument in favor of the first alternative.

Many similar arguments about molecular shape, molecular size, and how closely molecules could pack together at surfaces were settled by the Langmuir technique. Scientists, however, were still no closer to understanding what it was about detergent molecules that made them pack spontaneously into different structures. The question was especially important in the case of lecithin (the material that is sold in health food shops), since lecithin is a major component of biological cell membranes, and it had been found that lecithin spontaneously forms liposomes when shaken up in water. The exciting possibility was that the first cell membranes ever to occur on Earth might have been formed in this way, especially since it had been shown that lecithin molecules were probably present in the prebiotic soup.

Suddenly everyone, including myself, was trying to work out how lecithin molecules get together. My approach had to be indirect because by now I was working in a government food research laboratory and experiments on the origin of life were rather outside my remit. Lecithin was a food material, often used to hold water in place in the oily matrix of foods like margarine, so I tackled the question of how much water the lecithin could carry, although this was mostly a pretext for keeping in contact with what others were finding out about the aggregation of lecithin molecules to form complex structures.

They were finding out a lot, though not much that was of lasting value, partly because it was difficult to obtain pure lecithin. It was for this reason that some scientists decided to try materials that could be obtained in pure form and which would also form bilayers. One of those people was a Cambridge scientist named Denis Haydon. The materials that he tried were the *monoglycerides*, which are molecules formed by the breakdown of oils and fats as part of our natural digestive processes. I never got around to asking Denis just why he chose these materials, but it seems in retrospect that he had a glimmering of the revolution that was to come, and sought to use surface-active molecules whose shapes ensured that they would pack neatly into flat sheets. He knew that one monoglyceride in particular, called *glycerol monooleate*, or GMO, had the right dimensions to fit the bill, partly owing to some earlier work on monoglycerides by a group whose star student in the use of the Langmuir trough was a young grocer's daughter called Margaret Roberts, who was then embarking on a career as a food scientist. As it turned out, Margaret Roberts' first ˋ piece of scientific work was also her last. She married a man named Denis Thatcher, decided that politics was a far more exciting and rewarding occupation, and became Britain's first woman prime minister.

It turned out that GMO was an ideal material for making artificial bilayers. Denis Haydon made them by dissolving the GMO in oil and painting the solution over a hole in a piece of Teflon separating two halves of a small tank full of water. The oil drained away to the boundary, leaving a film whose structure was basically similar to that which occurs in natural membranes (Figure 7.9).

Such films became known as *black lipid membranes*, or *BLMs*,

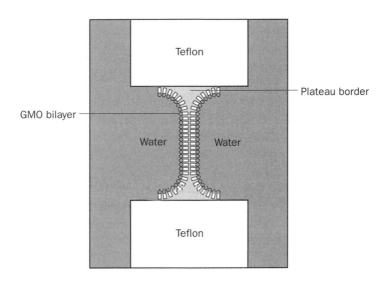

Figure 7.9: Schematic Representation of Glycerol Monooleate Bilayer.

because they were so thin that they reflected very little light. The boundary, which is the reservoir to which all of the excess solvent drains, is known as the Plateau border in honor of the Belgian Josef Plateau, whose fame among surface scientists comes partly from the fact that he worked out experimentally the elegant rules governing the angles between the corners and edges in a foam (which are also called Plateau borders), but still more from the fact that he did all of this after he had been blinded during experiments in which he had looked for too long at the sun. Despite this, his published papers are full of the most beautiful and accurate diagrams, produced with the aid of a sighted assistant.

Denis's work on black lipid membranes ultimately gave a good deal of information on the processes that guide self-assembly, and in particular on how other molecules can worm their way into bilayers without disrupting the whole structure. Many of the molecules that he studied were local anesthetics, and he discovered that these work by fattening the bilayer structure so that membrane molecules whose job it is to pass material from one side to the other can no longer span the full distance.

Among the other questions that Denis tackled was the vexed phenomenon of why the tails of detergent molecules prefer their own company to that of water. It was a case of "oil and water don't mix" — but *why?* He was unable to come up with an answer — in fact, no one was. Even now, we don't really know what it is that makes oil virtually insoluble in water. It's not the lack of attraction between the molecules — individual oil and water molecules can cohabit quite happily. It is more that water molecules prefer to arrange themselves in a flickering, evanescent array. An individual oil molecule penetrating this array forces the nearby water molecules to adopt a more permanent arrangement, and is about as welcome as a marriage celebrant in a *ménage à trois.*

At least Denis was able to measure the driving force that pushes oil and water apart, by measuring the work required to increase the area of contact between oil and water surfaces — in other words, the interfacial tension. It was data such as these, together with information on the repulsive forces between the head-groups, the shapes of different sorts of detergent molecule, and the ultimate packing density of different types of hydrophobic tail, that Jacob, Barry, and their colleague John Mitchell brought together in one beautiful synthesis that enabled them to predict just what sort of detergent molecules would form what sort of structures.

A Grand Synthesis

I was particularly interested in what they were doing because I had just written a review summarizing what was known about how molecules pack to form micelles. It had become abundantly clear that there was no way that most detergent molecules could pack to form a spherical micelle: there was simply no room in the middle for the ends of all the tails. There were plenty of suggestions about how this problem might be overcome, some involving very weird micelle shapes. My coauthor, David Oakenfull, and I favored the simplest, which was that micelles must have the shape of either a prolate or an oblate ellipsoid — in other words, they were either egg-shaped or discus-shaped — to allow room for the tails in the middle.

Even before our review appeared in print, Jacob, Barry, and John were at work on an idea that would leave our simple-minded arguments far behind. They argued that the only molecules capable of forming micelles would be those that could pack to form essentially spherical structures. As pointed out earlier by the American scientist Charles Tanford, molecules with other shapes would simply pack to form other structures. Jacob, Barry, and John refined this idea into a quantitative model that was complicated mathematically, but which led to one extraordinarily simple result. It turned out that the sort of structure into which detergent molecules pack depends on only three things: the length of the tail "L," the volume of the molecule "V," and the "optimal" cross-sectional area of the head-group "A" — in other words, the amount of space that the head-group prefers to have when in company with others of its kind. If $(V/(A \times L))$ is less than one-third, the molecules pack as spherical micelles. If it is between one-third and one-half, the micelles become ellipsoidal. If it is between a half and one, the molecules pack as bilayers or liposomes. If it is greater than one, inverted micelles are formed. The model was an immediate success, and a profusion of papers followed in which scientists around the world explained their results in terms of it.

One thing that the theory could explain was transparent *microemulsions* which, unlike normal emulsions such as milk and mayonnaise, were indefinitely stable. This was a difficult pill to swallow for scientists like myself, who had been brought up with the idea that emulsions, which are suspensions of droplets of one liquid in another (e.g., oil in water), would eventually "break" as the droplets floated to the surface and coalesced. We knew that we could slow the process down by coating the droplets with a protective layer of surface-active

material such as glycerol monooleate or the lactoglobulin protein with which milk fat globules are coated, but we also knew
that this was a holding measure at best, and that the oil and
water would eventually separate as the droplets coalesced.

Here, though, were new emulsions that seemed to be indefinitely stable. Their existence had been known for some forty
years, but the way in which they were stabilized was not understood until the "packing" theory was developed, when Barry
and others quickly realized that the secret lay in the shape of the
detergent molecules that formed them. Two sorts of molecule
with complementary shapes were used, so that the voids between one group were filled by the other. The result was an
emulsion containing extraordinarily tiny droplets — so small
that the emulsion appeared transparent. Some microemulsions
contained no droplets at all. The oil and water phases simply
wound in and out of each other in a series of intricate curves,
generating what came to be called a *bicontinuous* phase. The
mathematical theory of such systems permitted Barry and others to design new microemulsions. One of the consequences of
this was that Barry became an experimentalist, rather to Jacob's
disapproval. Another was that a range of new detergent products hit the market, with an attractive transparency of appearance and flow properties that let them sit as a gel until shaken,
when they promptly flowed at the user's command. A market
niche for these products was found in hair shampoos, for which
the detergents in microemulsions are ideal.

By this time I was 12,000 miles away, spending a sabbatical
year with Denis Haydon in Cambridge, pursuing an idea that I
had about black lipid membranes. My proposal was to use water pressure to bulge them up like balloons, so that they could
be pushed together in the same way that Jacob had pushed
mica surfaces together, but here modeling what happens
when two cell membranes come into contact. As it turned out,
I got no results for the whole year. In fact, it took a further
three years before I was able to bulge black lipid membranes
successfully and push them together, and a further two years
before I published the results of these experiments with Denis.

The results showed that the black lipid membranes jumped together from a distance of some thirty nanometers, immediately fusing to produce an unusual structure not normally
seen when two real biological membranes fuse. For this and
several other technical reasons the black lipid membranes
were not good models, and I abandoned this line of inquiry in
favor of working with real biological cells, which others were
finding contained a small proportion of molecules called
lysophospholipids that had relatively large heads and which
would normally be expected to occur as micelles. It is now believed that lysophospholipids are normally dispersed within
the membrane, but can get together when required to open
up a hole — the first step towards two membranes fusing to
become one in such processes as fertilization.

Was the work that I started with Denis wasted? In one sense,
yes, although it did lead to the development of a new experimental technique that is now finding application in such diverse fields as oil recovery and the manufacture of ice cream.
In another sense, no scientific effort is truly wasted, so long as
the question being asked is a serious one. The answers may
not prove relevant to the original question, but their consequences are largely unpredictable, as Pasteur, Franklin, and a
host of others have found by persistently pursuing particular
lines of inquiry. Persistence is one of the most valuable attrib-
utes that a scientist can have. Without the persistence of many
scientists, some going in the right direction, some concerned
with what may have appeared to be lost causes, we would
never have realized that the membranes that envelop living
cells are formed spontaneously by the "self-assembly" of many
small molecules, and that the very existence of living membranes depends critically on the shapes of those molecules.
This knowledge, applied in other areas, has led to the development of better hair shampoos, new foams for fire extinguish-
ers, and other practical applications. Applied to medicine, and
to the better understanding of ourselves and our evolution, it
is possible that it may even be the saving of the human race.

8
a question of taste

I was once asked to give an after-dinner speech at an important conference in Philadelphia on food tastes and aromas. On the night of the dinner I discovered that the after-dinner speech given the previous year was made by a Nobel Prize winner and that, on another occasion, the entertainment had been provided by a full dancing troupe from India. Any confidence that I might have had rather evaporated after I received this information, but at least I knew that I was going to be way ahead of the Indian dancing troupe in the popularity stakes. They had positioned themselves so as to block the path to the toilets, and then danced for half an hour longer than expected, just when the bladder-expanding effects of a particularly good chardonnay were being felt by the majority of those present.

My host, the chairman of the conference, had asked me to give a speech that was "informative but amusing." It was difficult to see what I could offer to an audience of world experts at a conference where even the dinner menu contained a detailed analysis of the flavor combination in each dish. If I tried to talk about flavor from the diner's point of view, then the amusement, I thought, would mostly be at my ignorance.

I thought that I could see a way, however, in looking at eating from the point of view of a physical scientist, and to concentrate on how foods release their tastes and aromas. I was reasonably well qualified for the task, having spent twenty years working as a physical scientist in a government research laboratory, and having spent quite a bit of time since working with chefs and food scientists. Even so, I found that there was a lot that I didn't know and, as I soon discovered, that no one else knew either, and I was forced to develop some new ideas that turned out to be novel to many present as well. Other scientists have since picked up on these ideas, or developed them independently, again showing how science is a community affair, and how you can never tell where something might lead. I have looked at eating from the point of view of the speech many times since — sometimes as serious science, sometimes using food as a hook to exemplify how scientists look at the world. Some of those stories are given in the following chapter, where I examine the science of eating a meal. All of the science is serious, even if not all of the stories are.

Introduction

A friend of the great nineteenth-century French gastronome Jean Anthelme Brillat-Savarin told the story of how he had visited the famous man in midafternoon, only to be kept waiting for some time. Eventually Brillat-Savarin appeared, full of apologies: "I was in the drawing room, enjoying my dinner."

"What?" said his guest, who knew that Brillat-Savarin *always* had his dinner served formally at the dining-room table. "Eating your dinner in the drawing room?"

"I must beg you to observe, Monsieur," replied Brillat-Savarin, "that I did not say that I was eating my dinner, but enjoying it. I had dined an hour before."

Brillat-Savarin was one of the first people to analyze the art of gastronomy. He summarized a lifetime spent largely at the meal table in a two-volume compendium published in 1826 with an unwieldly title that begins *Physiologie du goût, ou Méditations de gastronomie transcendante, ouvrage théorique, historique et à l'ordre du jour* (The Physiology of Taste, or Meditations on Transcendent Gastronomy, a Work Theoretical, Historical, and Programmed), which lists precepts, anecdotes, and observations on how to enhance the pleasures of the table. We still follow many of Brillat-Savarin's precepts, in principle if not always in practice. The science underlying those precepts, however, is only just now beginning to be understood, and many basic questions remain unanswered. Why, for example, does the visual appearance of a meal affect our perception of its flavor? How do taste, aroma, and other factors interact to produce the sensation that the French call *goût*, and for which the closest English word is *flavor*? How do foods release taste and aroma molecules that affect receptors in the tongue and the nose respectively to produce this sensation?

In this chapter I follow the science of eating a meal from start to finish, to examine the progress that scientists have made in understanding what happens as we look, chew, and swallow.

Perception

A good meal begins with expectation, continues with gratification, and ends with satisfaction. Brillat-Savarin was one of the first to recognize that the first of these, expectation, is an important part of the enjoyment of a meal. He believed that the expectation produced by an appropriate ambience was especially important, and that the enjoyment of a meal when dining out was enhanced by "an elegant room [and] smart waiters," as well as "a choice cellar, and superior cooking."

Modern psychologists would agree with Brillat-Savarin. They have shown that expectations based on ambience, lighting, the company present, the food's appearance, aroma, and texture, and even the quality of the table napkins, can all affect the way that food *actually* tastes to the diner. These effects are not confined to gourmets, or to those who imagine themselves to be gourmets. A commercial hamburger that tastes great to a teenager when eaten in the company of a group of friends may taste terrible to the same teenager when eaten in the company of his or her parents. A restaurant meal that melts in the mouth when shared with a lover will not taste quite the same to someone having a domestic argument with his or her spouse. The expression "the food turned to ashes in my mouth" has real meaning when it comes to the perception of flavor. We taste with our brains. All that our tongues and noses do is send sensory information to the brain about the taste and aroma molecules reaching them. The brain processes the information, together with whatever other information it perceives to be relevant, and produces a response.

The "other information" can come from some curious sources. The Italian Filippo Tommaso Marinetti reported in his *Futurist Cookbook* of 1932 the experiment of eating the same food while resting his fingers lightly on either velvet or sandpaper. The perceived texture of the food, he reported, was quite different in the two cases. Marinetti dined his way from Milan to Paris to Budapest, staging eye-catching demonstrations with his talks as he ate meals that included "Raw Meat Torn Apart by Trumpet Blasts," and a recipe that included chickpeas, capers, liqueur cherries, and fried potato chips, all eaten individually between carefully clocked stretches of silence. The Futurist movement "disdain[ed] the example and admonition of tradition in order to invent at any cost something new which everyone considers crazy." Their aim was to shock. Their modern successors are chefs such as Heston Blumenthal, chef-proprietor of the Fat Duck restaurant at Bray beside the River Thames near London, who aims to use the brain's responses to create new flavor experiences that might not shock, but certainly surprise.

Aromas, for example, usually contain hundreds, or even thousands, of different chemical compounds, and some aromas, such as garlic and coffee, have major components in common. Heston has tried mixing the two materials in a crème brulée. The mixture sounds horrible, but it seems to fool the brain, which can't decide whether it is experiencing garlic or coffee, and oscillates between the two, enjoying each one separately but never being able to "mix" them.

Aroma is not the only thing that appears to trick the brain. Heston also makes a beet jelly, to which he adds tartaric acid, the main component in the "crust" thrown by a good wine. Tartaric acid, as its name suggests, makes the jelly taste "tart," and this, combined with the visual appearance, gives the taster the impression that he or she is eating blackcurrant rather than beet. One taster, when told that the jelly was beet, said that it tasted disgusting. When told it was really blackcurrant, however, she decided that it was actually delicious. But it was really beet.

Heston bases much of his culinary art on the premise that the human brain loves surprises, a premise that has now been supported by scientific experiments. Surprises arise from a contrast between expectation and experience, but expectation can prove dominant in many cases. It starts with the appearance of the food on the plate, which can even affect whether the diner is prepared to put the food into his or her mouth. Most people have heard of experiments where meat is served under green or yellow lighting, with the result that diners cannot face eating the food. So far as I know, the experiments have not been extended to the effects of flickering TV light on the perception of a meal — it would be an interesting study.

Even if the lighting is normal, appearance can still have a dramatic effect on the acceptability of a food. When I was working in an Australian government food research laboratory, I was a member of a tasting panel whose job was to evaluate the flavor of vegetarian sausages, produced from soybean protein by a local agricultural college. The sausages were perfectly normal in appearance, a good crisp brown, and neatly presented on a white china plate with stainless steel cutlery. All was well until we applied our knives and forks to one end of the sausages, when the contents promptly ran up to the other end in a liquid mess. There was no way that I, or any other member of the taste panel, could put those sausages in our mouths.

The appearance and texture of food on the plate give important, though sometimes misleading, cues as to what we might expect when we put the food in our mouths. That we respond to these cues so strongly suggests that the response is deeply embedded in our psyche, perhaps from prehistoric times where crucial distinctions had to be made between foods and the positively dangerous. The really strong cue, though, is the one that is used by most animals, including ourselves — aroma.

There is nothing like the aroma of a fresh meal, and so far there is nothing to replace it. At a meeting I attended in Sicily, the French chef Anne-Marie DeGennes produced two ratatouilles, one which had been flavored by fresh thyme and bay leaf, and the other by extracts from these same plants. The first was undoubtedly preferable, even though no aroma compounds had been lost in the extraction process. The difference was probably due to the fact that the extraction procedure was more efficient for some aroma compounds than others, so that the balance in the complex mixture of aromas was subtly changed.

Even so, flavor scientists are making progress. A drop of the compound hexanal, which produces the "green note" characteristic of fresh fruit and vegetables, can restore that note of freshness to a cooked dish. Some may balk at adding a "chemical," but the world is made of chemicals, and hexanal is one that plants produce for themselves and which we eat all the time. Why not add it in pure form, rather than following the chef's procedure of throwing in a handful of fresh material, as they do when they add a purée of fresh asparagus tips to enhance the aroma of asparagus soup? Chefs often "cheat" in this way, if you can call it cheating. So do home cooks. Who has not heard of, and probably used, the trick of putting a few coffee beans under the grill so that the aroma will enhance the appeal of instant coffee? This trick has been the subject of psychological tests, where blindfolded subjects were given hot water to drink while being allowed to smell the aroma of fresh coffee. All of them were convinced that they were drinking the real thing.

We often enhance the aromatic appeal of our meals prior to ingestion by coating them with gravy and sauces. Because of their liquid constitution, these materials release their aromas more readily than do the foods that they coat, and provide an enhanced aroma experience and flavor expectation even before the food reaches the mouth. People quickly come to associate appearance and aroma, which is one of the reasons a rich brown gravy is so favored on roast meals, especially in England, the home of the roast dinner. The use of gravy is so common that one would think it had few surprises left to offer, but one would be wrong, as I found when I was asked by an advertising firm to investigate the science of gravy.

I only undertake such projects on rare occasions, if I can see them as a way of drawing public attention to the fact that science is everywhere around us, and not just confined to medical advances, the destruction of nations, or the fate of the universe. This project certainly drew public attention, to the extent that I received a delighted e-mail from Marc Abrahams, organizer of the IgNobel Prizes, saying that I might be in line for a second award. In the beginning, all I could think of doing was to look at how much gravy would be taken up by the various components of a traditional roast dinner. Meat, I thought, would obviously take up quite a bit, as would mashed potatoes, while roast potatoes, peas, and beans would absorb very little. As it turned out, I was almost 100 percent wrong.

I performed the experiments with the help of my colleague, Peter Barham, who did the necessary cooking in his home kitchen. My wife, Wendy, kept the experimental records and acted as umpire in cases of dispute. Our procedure was simple. We drew up a list of the most common ingredients in a traditional roast dinner, weighed them, cooked them separately, weighed them again, soaked them in the gravy specified by the advertisers, wiped off the excess gravy, and weighed them yet again. For good luck, we measured the dimensions of the food pieces before and after cooking, following the basic scientist's principle of measuring everything that might conceivably be relevant.

Some of the weight changes were only a fraction of a gram. It soon became apparent that Peter's kitchen scales were not sufficiently accurate, so we borrowed a top-loading balance that was normally used to weigh out small quantities of fine chemicals, cleaned off the more poisonous of the residues, and started again.

Our first surprise was that roast meat does not take up any gravy at all — none. The second was that mashed potato does not take up gravy either — in fact, mashed potato *loses* weight when dipped in gravy. There were more surprises to follow. Peas and beans, it turned out, take up quite a lot of gravy. So do roast vegetables (up to 30 percent in the case of roast potatoes). For some vegetables, such as parsnips, the amount of gravy absorbed turned out to depend on how the vegetable was cut before cooking. The amount of gravy taken up also seemed to depend on which end of the cooked vegetable was dipped in the gravy.

We put our results together in a table, where we included the traditional English Yorkshire pudding, a porous dough that takes up an incredible amount of gravy, and bread, which many people use to mop up excess gravy. The results are shown in Figure 8.1.

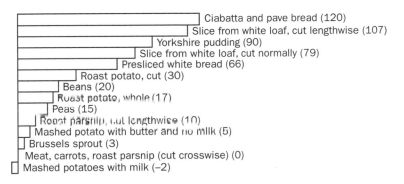

Ciabatta and pave bread (120)
Slice from white loaf, cut lengthwise (107)
Yorkshire pudding (90)
Slice from white loaf, cut normally (79)
Presliced white bread (66)
Roast potato, cut (30)
Beans (20)
Roast potato, whole (17)
Peas (15)
Roast parsnip, cut lengthwise (10)
Mashed potato with butter and no milk (5)
Brussels sprout (3)
Meat, carrots, roast parsnip (cut crosswise) (0)
Mashed potatoes with milk (−2)

Figure 8.1: Gravy Uptake Ratings (Percentage of Cooked Weight).

What were we to make of all this? The simplest picture that we could think of was that the moisture lost in cooking leaves a void volume that can be refilled with gravy (Figure 8.2). Was there some way that we could test this picture?

There was. We had weighed the materials before and after cooking, so we could work out the percentage moisture loss during cooking. If this was the same as the percentage weight gain when the cooked food was dipped in gravy, our model was proved, or at least extremely well supported.

As every cook knows, though, some foods shrink when they are cooked, reducing the space available for refilling by gravy. So we had to include a correction factor, calculated from the change in the physical dimensions of the food, to allow for this shrinkage effect. It was just as well that we had taken the appropriate measurements. The shrinkage factor was small for most foods, but for meat it turned out to account completely for the water loss, and there was no void volume left for the gravy to enter. That was one puzzle explained. Another puzzle concerned the peas and beans, which do not shrink when cooked — if anything, they swell. A closer look soon revealed the answer — the outer casing of peas becomes loose, and the seeds drop out of many beans, in both cases leaving void volumes that fill up when the vegetable is dipped in gravy. That was another puzzle explained. The third query concerned the mashed potatoes. This time we used the Hercule Poirot approach, and our little gray cells, fed by suitable wine, soon reminded us that potato mashed with a little milk is already totally saturated with liquid, and has no room for more.

With these problems out of the way, we sat down to compare the percentage moisture loss (corrected for shrinkage) with the percentage gravy uptake for the various foods that we had cooked. If our hypothesis was right, then the two quantities should have been the same for each food. They were, with a closeness of fit that would have pleased any experimental scientist, as Figure 8.2 shows.

Our model seemed pretty good, although there was still plenty left to chew over. Why, for example, did baked potatoes take up to ten minutes to absorb the maximum amount of gravy, while baked parsnips took less than a minute? Why did parsnips cut "across the grain" absorb no gravy at all after roasting, while those cut lengthwise absorbed up to 10 percent? Why did roasted vegetables take up more gravy through the side that had been in contact with the roasting pan?

These questions didn't bother the media, who were in-

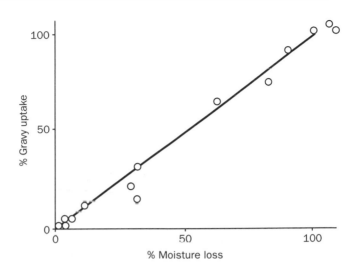

Figure 8.2: Gravy Uptake Ratings (Percentage of Dry Weight).

trigued by the idea that science had something to say about something as homely as gravy, and that a scientist had come up with a "Gravy Equation." They bothered me, though, because unanswered questions are always a bother to a scientist. The answers were unlikely to be useful to mainstream science, like Rumford's observations on apple pies, or Benjamin Franklin's insight after seeing ship's cooks toss their greasy water overboard. The best I could hope for was that some of the answers might be relevant to restaurant and cooking practice — in other words, the science of eating.

Chewing

The science of eating was defined by the satirist Ambrose Bierce in his *Devil's Dictionary* as "performing successively (and successfully) the acts of mastication, humectation and deglutition" — in other words, biting and chewing, mixing the chewed food with saliva, and swallowing the result. I once attended a meeting on the science of gastronomy where this whole process was displayed in a stomach-turning X-ray video that showed a shadowy skull, complete with spectacles, rhythmically chewing and eventually swallowing a bolus of food (Figure 8.3). Fortunately, not many of us get to view eating in this way, unless our eating partners are particularly open-mouthed. Once the food disappears into our own mouths, all that matters is the flavor, or *goût*.

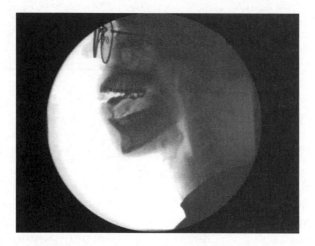

Figure 8.3: X-Ray Video of Chewing Head.

The person who showed the film was concerned with what happens when people experience problems in eating, and was using the film to show what "normal" eating looks like. One of his conclusions was that we tend to orient asymmetric pieces of food with the long edge parallel to the line of our teeth, and that the pieces of chewed material from the two ends do not mix. If the tastes at the two ends are different, as they will be on a slice of toast with, say, pâté on one end they will not mix initially as we chew. One will stay on the right side of the mouth, and the other will stay on the left. Clearly, the way in which we cut and present our food can affect the taste outcome, although as far as I know this aspect of food presentation has received little or no attention.

There are two main points to chewing. One is to break the food into pieces small enough to swallow. The other is to release tastes (experienced by the tongue) and aromas (experienced by the back of the nose), both of which contribute to the final flavor experience. The tastes and aromas, in turn, promote the flow of saliva, making the food easier to swallow.

The forces that go into breaking up the food are transmitted both by the teeth and the tongue, and are extraordinarily difficult to model satisfactorily; far more difficult than those involved in sending a rocket to the Moon, which has been modeled successfully on a number of occasions. By "model" I mean writing down equations (which may be fed into a computer) that will predict the course of events. The word can also have a practical meaning: the food industry is full of instruments that "model" how we eat our food. Most of these are instruments designed to measure texture. They range from the simple pea maturometer, which is nothing more than a spring-loaded pin that is stuck into peas to measure how hard they are, to complicated instruments that mimic the action of a chewing jaw. All of these instruments, without exception, are close to useless when it comes to predicting the texture, or "mouth-feel," of a food when it is actually eaten.

The problem, in physical terms, is that chewing is a highly "nonlinear" process: in other words, the effect is not proportional to the action. In the case of chewing, a very small change in the way we chew can have a huge effect on the outcome. A bite on a gingersnap with a force of 4.99 kilograms, for example, might fail to break the cookie, whereas a force of 5.00 kilograms may shatter it. It is difficult to study nonlinear processes, even in a laboratory, and still more difficult to translate the results into practical situations. This applies especially when the effects progress all the way down the line, as they do in chewing, where breaking up food particles is only the first step. As they are broken up, the particles gather together in a lump called a *bolus*, whose response to the forces exerted by the teeth and tongue is also likely to be nonlinear. The bolus may, for example, be "shear-thickening," becoming stiffer as it is chewed on, or "shear-thinning," where the more it is chewed, the more liquid-like it becomes.

The normal bolus can show both behaviors, a fact that seems to have been known to the nineteenth-century British prime minister Sir William Gladstone. Gladstone, a careful and patient man, recommended chewing each mouthful of food thirty times. As it turns out, the bolus is normally shear-thickening over this range, reaching its maximum cohesion after thirty or so chews, a point which the body takes as a signal to swallow. With more chewing, the bolus becomes shear-thinning and eventually breaks up. A further complication in the analysis of the bolus is the presence of saliva, which makes solid pieces of food easier to break up. Gravy and sauces have a similar effect, which arises because the presence of a liquid makes it easier to stretch, bend, and open cracks in a solid surface.

Overall, we know a lot less about what happens during chewing than we do about what happens in the interiors of stars. One thing that we do know, though, is that chewing works, not just to break food into manageable bits, but also to release its flavors. This is an area we are now learning quite a lot about, especially when it comes to understanding what happens when taste and aroma molecules reach the tongue and the nose respectively.

Flavor — A Mixture of Taste, Aroma, and Pain

The molecules that are released from the food bolus in the process of chewing induce three main sensations — taste, smell, and, surprisingly, pain.

Taste happens, obviously enough, on the tongue and palate, which can distinguish five basic taste sensations — bitter, sweet, salt, sour, and "umami," described as "meaty, brothy, full-flavored." Bitter tastes are experienced because the molecules that induce them bind to particular protein molecules (called receptors) in a sensory cell's membrane. The sensory cells are gathered together in taste buds, which are housed in groups of three to fifteen in the tiny but visible bumps on the tongue called papillae. When a molecule binds to a receptor (often because the molecule has a shape that allows it to fit as a "key" into the receptor's "lock"), the receptor protein passes a chemical message to the inside of the cell to say that binding has occurred. The cell's response is to emit an electrical signal that is sent to the brain, which registers, "Ah, bitter!" Bitter tastes are there for a reason: to warn us against eating a particular food, often because it is poisonous ("bitter almonds," for example, contain traces of cyanide). It is surprising, then, that some of our foods of choice (such as beer and dark chocolate) owe much of their appeal to bitterness, and there is no real explanation why this should be so, although it is noteworthy that the majority of "preferred" bitter foods are pharmacologically active. We have some sixty different receptors for compounds that taste bitter; when a person is missing one of these receptors, they become "bitter-blind" to the compounds that it responds to. Around a quarter of the population, for example, is genetically bitter-blind to the compound phenylthiocarbamide, which is detected as incredibly bitter by the remaining 75 percent. Another quarter of the population (mostly women) seem to be bitterness *supertasters*, able to detect bitter compounds at abnormally low levels. In these people the papillae are tightly clustered and surrounded by a unique ring structure, whose function is not yet known.

There are taste buds that respond to bitterness all over the tongue, although most are concentrated towards the back. Taste buds that respond to sweetness are concentrated towards the front, which is hardly surprising as it is this part of the tongue that first encounters the lactose (one of the sweetest of sugars) in mother's milk. Curiously, "sweet-blindness" is practically unknown, perhaps because a taste for sweetness is a survival trait.

The other taste that uses a receptor is the controversial

"umami," a taste described variously as "brothy" or "meaty," and which I have named the "Wow!" factor in a newspaper column. It is induced by monosodium glutamate, the well-known MSG that is frequently added to Chinese food in particular as a flavor enhancer. MSG occurs naturally in many foods, including tomatoes and Parmesan cheese. It exists in the cheese as visible white crystals that can be dissolved in water, leaving a cheese distinctly lacking in flavor.

Umami has been claimed to be a separate taste because a receptor for glutamate has been discovered on the tongue, but there is still some argument whether MSG (and some other compounds that also bind to the receptor) itself has a separate taste or not. Perhaps the choice itself is a matter of taste. The two other tastes — saltiness and sourness — use different types of receptor and are general effects of acid (for sourness) or salt on the tongue as a whole.

Tastes can affect each other in surprising ways. One is the well-known enhancing effect of a small amount of salt on the perception of sweetness, which is why recipes for sweet scones often specify the addition of a pinch of salt. Salt can also affect the perception of bitterness. I once participated in a tasting of red wine where we were asked to assess the bitterness caused by the presence of tannins, which are universal components of red wine. We were then given some salty crackers to eat, and asked to repeat the test. The bitterness of the wine was definitely diminished, which is perhaps one reason why red wine goes well with savory food.

Bitter foods can saturate the bitterness receptors and thus wipe out the bitterness of a subsequently eaten food, which is why chefs avoid such combinations or progressions. Try drinking some tonic water before eating your next piece of dark chocolate. The bitterness of the chocolate will be eliminated, leaving only the taste and the mouth-feel of wax.

There is also a well-established relationship between a liking for sweet things and a liking for alcohol. Both of these materials seem to activate the same nerve signals to the brain. Conversely, an induced aversion to sweet things can induce an aversion to alcohol, although so far as I know this effect has so far only been observed in mice.

Taste is complicated enough, but the complications multiply when it comes to aroma, where there are over 3,000 types of receptor to be considered, 2,000 of which are active in humans. We have most of these receptors in common, but hardly any of us has exactly the same set. Even for the limited range of odors so far studied, there are some for which only 10 percent of the population have a receptor, others for which around 1 percent of the population are missing a receptor, and some twenty or thirty odors in between these two extremes. It is likely, then, that each of us has an experience that is a little, or even a lot, different when it comes to picking up the aroma of a meal. A case in point is the famous truffle, described by Rossini as the Mozart of fungi, where some 40 percent of the population are "tone-deaf" to the central aroma that gourmets rave about. Even among those who can detect this aroma, the effect ranges from sensual arousal to outright repulsion.

Aroma is generally more important than taste when it comes to flavor perception. This point can be shown very clearly by cutting an apple and an onion into equal-size cubes, and having a friend feed them to you in random order while you pinch your nose shut and close your eyes. Without the aroma and visual cues, most people find it impossible to distinguish between the two foods, even though their flavors are very different. Even with aroma cues, people cannot always distinguish between apparently dissimilar foods. Gruyere cheese and honey, for example, are very difficult to distinguish by aroma alone.

Aroma reaches us in one of two ways. The first is from the outside, as when we initially detect the smell of a meal. The second is from the inside, when the aroma reaches the back of the nose *retronasally* — that is, via the back of the mouth and nose after we have taken the food into our mouths. Aromas that reach us from the outside seem much more powerful when we sniff, but it is only recently that a group of scientists, led by Andy Taylor at the University of Nottingham, have found out why. The answer is not that we take in more of the aroma, but that we take it in at a more rapidly varying rate. It is the rate at which aroma concentration changes in the nose, rather than the concentration itself, that determines how strong we perceive the aroma to be.

The same effect happens from the inside. This was only discovered through the use of a clever piece of technology developed by Andy and his team called "MSNose," although I prefer the slightly more surreal name of "NoseSpace." The technique analyzes food aromas released to the back of the nose during chewing and measures their concentration in the nose in real time (Figure 8.4).

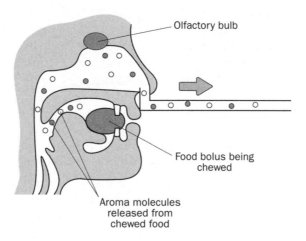

Figure 8.4: MSNose.
Aroma molecules from chewed food reach the olfactory bulb (where smelling takes place) retronasally, i.e., from the back of the mouth. MSNose takes samples of each breath to analyze the aroma molecules that have reached the back of the nose at that time.

It does this by taking a sample of each breath as the subject chews. The sample is taken via a tube inserted (quite painlessly) into the subject's nose. This hardly creates ideal conditions for gourmet enjoyment, but it is all worthwhile in the cause of science, as I discovered when I used it to study the effect of dunking a cookie on flavor release. This project was a follow-up to the original "cookie-dunking" exercise. The premise, or hypothesis, that we were testing was that dunking improves cookie flavor, and that it does so because the drink helps to release aroma molecules from the solid matrix of the cookie. Hot tea, we expected, would give the most release by warming up the volatile aromatic materials. For completeness, though, we tried cold tea, hot and cold chocolate, hot and cold milk, and even orange juice. As things turned out, it was just as well that we did.

My guide in the use of MSNose for these experiments was Rob Linforth, who had been happily unemployed until he joined Andy's group as a result of a casual chat outside the local newsdealer, and who now directed the use of the technique. Rob sports a magnificent, foot-long red beard, symmetrically divided to reveal the fact that he seldom wears a tie, and was ideal photogenic material for the advertisers who were supporting the project. The only person who didn't think so was Rob, whose excellence as a scientist is matched only by his dislike of cameras, especially when they interfere with the smooth running of his laboratory.

There was very little time to use the MSNose in between more serious research applications, but thanks to Rob's expertise we managed it, and an experiment that should have taken weeks was completed in just one day, albeit with the odd shortcut. Rob kindly sacrificed that day, although I felt at the time that my own sacrifice was just as great, since I had to dunk and eat 140 cookies with a stainless steel tube thrust up my nose, chewing for a prescribed count while Rob pressed buttons and recorded and analyzed the results.

The outcome was very surprising. Hot tea, it turned out, had no effect whatsoever on the concentration of aroma molecules released from the cookie and reaching the back of the nose. We later worked out that this was because the aroma molecules were washed down the throat before they had a chance to reach the nose. The drinks that worked really well were cold milk and, better again, cold chocolate. It was easier to find

a retrospective explanation than it had been to predict this result. The explanation I eventually came up with was based on
the fact that most aroma molecules dissolve easily in fats and
oils, but not in water (which is one reason why "fat-free"
foods tend to be so dull to eat). Milk contains fatty globules,
which absorb aromas from the dunked cookie and which coat
the tongue and tend to linger in the mouth when the dunked
cookie is eaten. These globules will gradually release aromas
that they have absorbed from the cookie, even though the
cookie itself has long since disappeared down the throat.
Chocolate milk, on this hypothesis, should be even more effective, because it contains cocoa solids, which are fatty and
which will also absorb aroma molecules from the cookie. Such
solids also hang around in the mouth (just look in a child's
mouth after he or she has had a chocolate drink), and will
gradually release aromas absorbed from a dunked cookie.

That was the hypothesis, which we were unable to test further because the machine had to be returned to other activities. Nevertheless, it made sense, and turned out to be a great
hit with the media, who turned up in droves for a series of
photo sessions, much to Rob's dismay. One of those sessions, a
live broadcast on CBS television's *Early Show,* provided inci-
dental evidence of the extent to which scientists travel when a
friend from the next-door laboratory back home rang to say
that he had seen me on TV in his New York hotel room the
previous morning.

MSNose has revealed unexpected facets of the way that we
detect and interpret aromas. It has shown, for example, that
many of the aromas released by a tomato when we eat it are
not present in the original tomato. The aromas that our brains
interpret as flavor are produced mainly once the tomato is
damaged by cutting or biting, processes that burst the cells
within the tomato, permitting the contents to mix and react
chemically.

Another success of MSNose concerns the effect of sweetness
on the perception of mint flavor. Most people will have noticed that a minty chewing gum slowly loses its flavor as it is

chewed. Gum chewers find that the flavor can be refreshed by taking a sip of a sweet drink. Experiments with MSNose have shown that the concentration of minty aroma molecules in the nose is unchanged throughout the whole process. It seems that the nose gradually becomes accustomed to the presence of a minty aroma, and its response becomes dulled. Sugar, which has no direct effect on the nose, nevertheless stimulates the brain to "notice" that there are still signals coming in which say that there is a minty aroma present. This is an extraordinarily clear case of how one sense can affect another. There are many others. Menthol, for example, the active component of minty sweets, can modulate our perceptions of hot or cold. A menthol solution in warm water will feel hotter in the mouth than water at the same temperature. Conversely, a menthol solution in cold water will feel *colder* than water at the same temperature.

Some "pure" odors, such as pinene (pine odor) and cadinene (juniper, an important component of gin), can even produce a sense of pain. The nerves concerned are called the *trigeminal* nerves. It is through these that we feel the acute pain of a blocked sinus. When we sniff pinene, cadinene, or a small range of other odors, the trigeminal nerves are stimulated. An oddity of this effect, as carefully controlled experiments have revealed, is that we can't locate which nostril the odor is coming from.

That's not so much of a problem when we eat food containing hot chili peppers, where the pain comes instead from the tongue. The hotness of chili peppers is due to a family of tasteless, odorless chemical compounds called capsaicinoids, which are manufactured and stored in the seed-producing glands along the ribs of the fruit. The capsaicinoid molecules bind tightly to the surfaces of trigeminal cells in the mouth, nose, throat, and stomach, whose normal job is to warn the brain of pain-causing damage. Capsaicinoid binding causes the cells to send the same message, even though there is no actual damage. This is what the chili pepper addict has been waiting for. According to one unproven theory, the brain responds by releasing endorphins, natural painkillers that create a feeling of euphoria in the absence of pain (the source of "runner's high"). An alternative theory, also unproven, ascribes the attraction of chili peppers to "benign masochism," like the thrill of riding on a roller-coaster, where the rider can experience feelings of fear in safety, knowing that all will be OK in the end.

How Does Chewing Release the Flavor?

The effect that a molecule produces, whether it is pain, aroma, or taste, can only occur if the molecule is released from the food and can reach receptors or other sites in the mouth and nose. How does this happen? This question, obviously an important one for gastronomy, was the central theme of my speech in Philadelphia, but it is not a question that is often asked. When I began hunting around in the relevant literature, I found that the few answers posited had not stood up to experimental test. The common picture seemed to be that aroma molecules diffuse out of the food bolus and are then swept by air currents (caused by breathing) to the back of the nose. The problem with this picture, it seemed to me, is that the bolus is generally a pretty watery mix, whereas most aroma molecules dissolve in oils or fats, which are likely to be dispersed as droplets or particles in the bolus. I pointed out to my audience that this leaves the aroma molecules with two problems. The first is to get to the surface of the droplet. The second is to get across the water barrier into the air space above.

I already had some information about the first problem through experiments that I had done with chef Fritz Blank in Erice. We had made a flavored mayonnaise in two ways: one was gently stirred and the other was beaten vigorously in order to make the oil droplets smaller. The flavor of the second mayonnaise came through much more strongly to tasters, partly because the droplets were smaller and so the flavor molecules took less time to reach the surface by diffusion, and partly because there was a much larger surface area for the molecules to escape through.

When William Gladstone claimed that he chewed every mouthful thirty times, he was probably doing the right thing according to this experiment, since chewing will tend to break the oily constituents in the food into smaller and smaller droplets, making it easier for aroma molecules to escape. They still have to make their way across a wide watery expanse, however. It was not clear to me how this could happen until I remembered some experiments that I had done years previously in connection with an entirely different problem, that of mineral flotation in mining. In this technique the ore is crushed, placed in a vat of water and detergent, and bubbles are blown up through it. The bubbles capture the desirable mineral particles in the ore and carry them to the top of the tank, leaving unwanted materials such as quartz behind. My task was to investigate how the particles are selectively captured.

It turned out that the surfaces of the mineral particles are oil-like (the technical term is *hydrophobic*), and when an air bubble approaches them the film of water in between the particles suddenly collapses at a thickness of a micrometer or so, allowing the air bubble to stick to the solid surface. "Could the aqueous film separating an oil drop from the air in a food bolus collapse in the same way when an oil drop gets within a micrometer or so of the surface?" I wondered out loud to my audience. "Could the thin film of water between the droplet and the air suddenly burst to let the droplet and its cargo of aroma molecules out into the air space above?"

The aqueous film contains more than just water, of course — it is full of salts, proteins from the food, and carbohydrates both from the food and the saliva. The presence of these materials may affect the chance of a water film bursting; but nobody knows, because nobody, to my knowledge, has looked. I speculated to my audience, some of whom were still eating, that food aromas are released in lumps, a droplet-full at a time as each droplet gets close enough to the surface of the bolus for the water film between it and the air to burst (Figure

8.5). Aroma molecules could then diffuse rapidly out of the droplets, providing aroma bursts to the nose, which is just what it needs to get the full effect, as Andy Taylor and his group discovered. As I made this suggestion, I noticed that the tardy diners had begun to chew their food more vigorously.

My host, Gary Beauchamp, director of the Monell Chemical Senses Institute in Philadelphia, pointed out after the talk that the mechanism that I had suggested also allows for a new possibility, which is that the droplets themselves, and not just molecules diffusing out of them, might be carried on air currents and reach the nose as an aerosol, bringing with them not only volatile aroma compounds but also involatile compounds that could not reach the nose by any other route, but which might in this way play a part in our aroma experience.

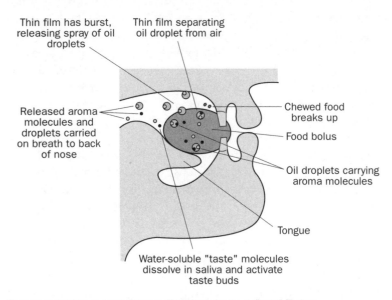

Thin film has burst, releasing spray of oil droplets

Thin film separating oil droplet from air

Released aroma molecules and droplets carried on breath to back of nose

Chewed food breaks up

Food bolus

Oil droplets carrying aroma molecules

Tongue

Water-soluble "taste" molecules dissolve in saliva and activate taste buds

Figure 8.5: Taste and Aroma Release from a Food Bolus.

A model such as the one above is only the first step in the scientific process. The picture that it portrays may sound convincing, but there have been plenty of models that have sounded even more convincing which have turned out to be wrong. These range from Aristotle's notion that objects only move if they are being pushed or pulled, to Thompson's "plum pudding" model of the atom. Each of these was believed in its day, because the model made sense of so many observations. But testing eventually proved that each model was wrong.

Much of the speculative "science" that appears in the media these days is no more than the presentation of a model developed to account for some set of known facts. No reputable scientific journal would accept such a model unless the author had made an effort to check it out in some way, usually by testing its predictions against reality. I could test out my model for aroma release, for example, by stirring a food bolus and monitoring the space above for a spray of oil droplets. I could also pass a current of air over a stirred food bolus, pass the air directly into MSNose, and watch to see if the aromas arrive in bursts. I have not yet done either of these things, and without such testing my model remains just that — a model.

People outside science often think that models, such as that for the behavior of light which Einstein proposed in his Special Theory of Relativity, are the *sine qua non* of science, the ultimate output of the scientific mind. It does not take an act of genius, though, to devise most scientific models — just a little imagination. That imagination may be fed by prior knowledge, insight, or even the memory of a dream. In a few cases, the models may even be a genuine product of genius. The true worth of any model, though, is that it is able to pass rigorous testing. Without this, the model is worthless, no matter how convincing it may appear to its creator or to others. The theory of relativity, however inspired, would be valueless if it were not for the fact that its predictions have turned out to be true.

On a rather lower plane, my picture of how food aromas are released will only have value if it turns out to be true. Nor did it need me to develop it, although as far as I know I was the first to enunciate it in public. Six months later, I mentioned its outlines informally at another food conference, and was promptly accused (tongue-in-cheek) of industrial spying by one of the other participants, who had, as it turned out, developed a very similar picture quite independently. It often happens that way in science, which is why linear "histories" of how one discovery led to another are so misleading. There is a community of people involved in any discovery, and most of the members of that community will be well aware, at least at a subliminal level, of what the major unanswered questions are, and will have put some thought into what the answers might be.

My picture of aroma release is now out in the open, as any piece of science should be, to be thought about, tested, dissected, and used if others so wish as a starting point for further work. I may or may not do some of that work myself — that depends, as detective-story writers say, on motive and opportunity. Even if the picture turns out to be true, it may or may not be important in the overall scheme of things gastronomical. Like the presentation of food on a plate, it is only a first step, and is yet to be chewed over, swallowed, and digested.

9
the physics of sex

I was once asked to give a talk to a school science club on any subject that I chose. I suggested "The Physics of Sex," a topic on which I had recently written an article, and was a little taken aback when the organizer agreed. I was told later that the audience had been rather larger than most meetings of the science club, and was also unique in that there were more teachers than students present.

Even the senior students were surprisingly vague in their knowledge of the physical aspects of sex. The highlight for me came at the end of the talk, when an earnest adolescent wanted to ask a private question. "Is it true," he asked in a conspiratorial whisper, "that if the girl is a virgin it goes bang?"

I don't know how he came by this curious belief; but at least it was a relatively harmless one. Other teenage beliefs may have more serious consequences. A survey of British general practitioners in April 2001 revealed that current adolescent thinking maintained that "Putting a watch around your penis before sex means the radioactivity of the dial kills off sperm." Other teenagers believe that standing on a telephone book during sex prevents conception, presumably because of some imagined effect of gravity on the sperm. This is a belief that goes back to Aristotle, who thought that males were conceived on the right side of the womb, and females on the left. He therefore recommended that a woman should lie on her right side after intercourse if she wanted a male baby, so that the sperm would drain in this direction.

Aristotle's beliefs were transmitted in a curious hotchpotch of information and misinformation for midwives called *Aristotle's Complete Masterpiece,* which appeared during the seventeenth century and was the most widely used source of information about sex in the English-speaking world up until the end of the nineteenth century. Its origin is unknown — only a little of the information in it can be traced back to Aristotle. It was banned in England for a long time on account of the explicit nature of its illustrations. When I bought my own copy, I found a newspaper cutting from the 1930s inside it with a reader's question, "Where may I buy a copy of *Aristotle's Complete Masterpiece,* and how much may I expect to pay?" The answer was a model of pragmatism: "You may not buy a copy of *Aristotle's Complete Masterpiece.* You may expect to pay three-and-sixpence."

Sex is even now regarded as a somewhat dubious topic for a scientist to be discussing outside a medical setting. I found this out when I was asked to give a predinner talk for a scientific society, and chose, once again, "The Physics of Sex" as my theme. Most people seemed to enjoy the talk, but one member of the society's executive sat through it with an air of obvious and deepening disapproval. It turned out that he had been led by my title to think that I was going to talk about the role of women in physics.

What I did talk about was the physical problems that a sperm has to overcome in the race to the egg, which involve diving, tunneling, surfing, and even synchronized swimming. At every stage, from the rocket-propelled launch to the final construction of an electrically guarded rampart, the sperm cell's journey is a model of the realpolitik of physics, and the winning sperm is the one with the greatest mastery of physics. The odds for any particular sperm's success are about the same as winning the lottery, but the reward is incalculable — life itself.

Step 1: Preparing for the Launch

Mammalian species such as man use the rocket launcher to propel spermatozoa on the first stage of their journey. Conventional wisdom decrees that the launcher must be erect and rigid to perform its function, but there is always someone who has to be different. One rebel, a forerunner of today's sensitive new-age guys, was an eccentric Australian who decided in the 1930s that erections were an unnecessarily forceful way of inserting sperm into a female partner, and that men of refinement ought to be content to permit the female partner to draw the flaccid penis in. Not content with maintaining this worthy policy as a private individual, he insisted on proclaiming it publicly every Sunday in the Sydney Domain (a public space reserved for soapbox declamations). In the climate of the day this was nothing less than pornographic, and the poor man was duly hauled off to the lockup each week, to be fined on the Monday and released to try again the following Sunday. Possibly the policemen who arrested him were aghast at what might happen if their wives heard of his arguments and took them seriously. At any rate, his arguments have disappeared into oblivion, and the rocket launcher continues to be used in the erect position.

The erection is a matter of *hydrostatics,* the branch of physics concerned with fluid pressure and the application of that pressure in the right place. The pressure in this case is blood pressure, generated by the pumping of the heart to push blood out through the arteries and have it return through the veins. In the penis, the arteries lie *across* the veins. Hormones released during sexual excitement relax the smooth muscle in the artery walls. The distended arteries press down on the veins and stop the blood that is entering the penis from escaping. The result is an erection, maintained by hydrostatic pressure. You can get the same effect by turning on a hose after putting a crimp in it or blocking off the nozzle.

When the pressure cannot be maintained, the result is impotence, a condition that is rumored to have been suffered by Henry VIII in his relationship with Anne of Cleves. It is a condition that I am assured by my medical friends can nowadays almost always be cured. One nineteenth-century "cure" was a splint made from bamboo and fitted over the penis, a procedure that must have been incredibly uncomfortable for both parties. Nowadays a simpler and more comfortable first approach is the brief application of a small vacuum device to induce a "passive erection."

For most men, the libido is a sufficient driving force to produce an erection. Some people are never satisfied, though, and have sought ways to enhance the libido by the use of aphrodisiacs. Unluckily for these people, there is no such thing as an aphrodisiac substance — the only real aphrodisiac is in the mind. That hasn't stopped people of both sexes from trying such things as tripe, rhubarb, and the necks of snails in the search for enhanced sexual gratification. These and other equally odd materials come under the heading of "sympathetic medicine" — in other words, they bear a fancied resemblance either to a penis (e.g., asparagus, ginseng) or to a vulva (e.g., oysters).

Some so-called aphrodisiac materials act as irritants to the sensitive mucous membranes. One such example is the nettle, which the Roman author Pliny suggested rubbing on the penises of underperforming bulls. This may explain why the ancient Romans were such fast runners. The best-known material in the irritant class, though, is *cantharidin*, the active principle in Spanish fly, i.e., blister beetles of the *Cantharis* or *Mylabris* genus. Cantharidin, formerly listed as an aphrodisiac by some commercial drug companies, was classified as a Schedule I poison on its last appearance in the British Pharmacopoeia of 1953. It is a very dangerous material, with a toxic dose of three milligrams and a fatal dose of thirty milligrams. Its effects include dry mouth, gastric pain, blood in the urine, and, eventually, death following kidney failure. The irritant effect of cantharidin may not enhance the libido, but at just the right dose it can produce erections. In a famous case in 1869, several *battalions* of French troops in North Africa reported to their medical officer with gastric pains and permanent erections. It transpired that they had been eating the legs of the local frogs, which had been feeding on the blister beetles prevalent in the area.

The British answer to cantharidin was cocoa. In the early 1950s sales of cocoa shot up when a rumor went around that it had libido-enhancing properties. The *New Statesman* magazine ran a competition in 1953 for the best poem celebrating this myth. The winning entry, "Cupid's Nightcap," was written by the appropriately named Stanley J. Sharpless:

> Half-past nine — high time for supper;
> "Cocoa, love?," "Of course, my dear."
> Helen thinks it quite delicious,
> John prefers it now to beer.
> Knocking back the sepia potion,
> Hubby winks, says, "Who's for bed?"
> "Shan't be long," says Helen softly,
> Cheeks a faintly flushing red.
> For they've stumbled on the secret
> Of a love that never wanes.
> Rapt beneath the tumbled bedclothes,
> Cocoa coursing through their veins.

Cocoa, like many "aphrodisiacs," was thought to have an equal effect on both sexes. One substance which does actually have an effect on both men and women, albeit in a negative sense, is alcohol, a vasodilator that may relax the inhibitions, but which unfortunately relaxes other things as well on the male side of the equation — hence the expression "brewer's droop." On the female side, it has been found by a substantial proportion of women to produce dryness and discomfort.

No drug is yet known that can excite the libido, but there are quite a few that can help to produce and maintain an erection that the libido has failed to stimulate. One of the earliest was *papavarin*, whose effects were spectacularly demonstrated by the psychiatrist Charles Brindley at a meeting of the British Andrology Society. Brindley, a born showman, began his talk by lowering his trousers, injecting a solution of the drug into his thigh, and "displaying the results throughout the duration of his hour-long talk." The exact dose of papavarin is critical, and cases have been recorded of people having to cut short holidays and return in pain with the same erection that they started with four days earlier. The idea of having an injection as a prelude to sex is also not one that many people would enjoy. Most people would prefer a stimulant that could be taken orally. It is this, rather than any aphrodisiac effect, that is the main advantage of Viagra (sildenafil citrate), a substance that can make even wilting flowers stand up straight.

Viagra was originally developed by Pfizer Pharmaceuticals as a drug for treating angina. According to one report, its dramatic effects on the rigidity of the male penis were only discovered after someone wondered why all of the male participants in the experiment had failed to return leftover pills once the trial finished. Viagra received the approval of the U.S. Food and Drug Administration as a treatment for impotence early in 1998, and has already built up a considerable folklore. One French restaurateur even developed a beef piccata in Viagra sauce, infused with fig vinegar and herbs. The creator of this dish, Jean-Louis Galland, said that he wanted to make his customers happy, particularly grandfathers and their wives. Unfortunately for Galland and his customers, the French authorities decided that he was dispensing drugs without a license.

Step 2: The Race Begins

Muscle spasms within the erect penis eventually launch the sperm on its journey. The ejaculate has been through a similar process to that of gasoline in a modern gas pump, where various additives are incorporated as the gas proceeds through the pump, until the final mixture that emerges is ready for action.

The manufacture of sperm cells (spermatogenesis) occurs in the testes, within the seminiferous tubules. The cells are transported through the epididymus, where they become motile, and then through the vas deferens in a solution of salts and proteins to their point of projection. The solution in which they are carried is enriched from the seminal vesicles by an alkaline yellow fluid called the seminal plasma. This solution contains hormones, enzymes, and metabolites, many of which have an unknown function.

The seminal plasma that is eventually ejected contains some 200–400 million wriggling spermatozoa. The odds that one of these will reach and fertilize the egg in any particular month are roughly 3:1 against, which means that a normal couple have a 90 percent chance of success in twelve months of trying. The odds of one particular sperm winning the race, though, are rather worse than the odds of a particular person winning the lottery with one ticket at the first attempt.

Human spermatozoa are about sixty micrometers long, with a flattish wedge-shaped head like a mini-surfboard. They must swim a thousand times their own body length to reach the egg — equivalent to a human swimming 1500 meters. For at least half of that distance, the sperm must swim through a material with the consistency of a thin jelly. First, though, it must escape from the seminal plasma that has carried it to the starting point of the race. Then, it must break into the jelly-like material, which is called *cervical mucus*. Nature has conspired to make sure that neither job is at all easy.

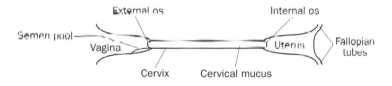

Figure 9.1: A Physicist's View of the Vagina, Cervix, Uterus, and Fallopian Tubes — An Obstacle Course for Spermatozoa.

The first obstacle is the seminal plasma itself, which is inimical to the health of the spermatozoa trapped in it, and will kill them unless they escape within twenty minutes. The semen settles as a pool near the entrance to the cervix, with the volume of the semen pool averaging three milliliters. Simple geometry says that the furthest distance that a spermatozoon needs to swim to get to the surface of such a pool is around nine millimeters. Most spermatozoa can swim at around three millimeters per minute in watery fluids, and so should make it to the surface in three minutes or so. Unfortunately for those optimistic spermatozoa, the seminal plasma, which travels through the penis in liquid form, sets instantly to a jelly on emergence. Only the luckiest or most vigorous spermatozoa make it out of this jelly. If the spermatozoa are sufficiently close to each other, they may display "synchronized swimming" — a hydrodynamic effect where the "waves" created by one spermatozoon affect the motion of closely adjacent spermatozoa, so that they end up swimming in unison. This phenomenon is used to assess the quality of bull semen.

Those spermatozoa that escape from the seminal fluid are immediately confronted with another barrier — the column of mucus that fills the cervical canal linking the vagina and the uterus (Figure 9.1). This extends between the internal and external *os* — a word of equal use to gynecologists and Scrabble players alike. The consistency of the mucus varies under the influence of the two hormones progesterone and estrogen. Progesterone makes the mucus more viscous, while estrogen induces it to take up water and become less viscous. The balance between these two hormones changes throughout the monthly cycle, and the consistency of the cervical mucus changes with it, becoming most easily penetrable near midcycle, although there is always a chance of some sperm finding their way in at any stage of the cycle.

There are various tests for the "goodness" of cervical mucus, i.e., the ease with which spermatozoa can penetrate and swim through it. The simplest, called the Billings test, is the one used in "natural" family planning, where a woman takes a small amount of her own cervical mucus between thumb and forefinger and measures how far it can be extended without breaking. The further it stretches, the better it is, from the point of view of the chance of conception. The degree of stretching seems to correlate quite well with the concentration of water in the mucus, and is easy to measure without recourse to complicated equipment. It is technically known as *spinnbarkeit*, for which the World Health Organization gives the following ratings:

Table 9.1: Stretch Ratings for Cervical Muous.

Stretching length (cm)	Rating
1	0 (worst)
1 – 4	1
5 – 8	2
> 9	3 (best)

If a microscope is available, it is also possible to measure "ferning," where some mucus is placed on a microscope slide, allowed to dry, and the resulting crystalline deposit is examined under a microscope. The more "branches" that the feathery crystals have, the better the mucus.

The most complete test reproduces the initial stages of the fertilization process *in vitro* — literally "in glass." Semen and cervical mucus are introduced to each other as thin films trapped between a glass microscope slide and a glass coverslip. Sufficient cervical mucus is placed in the gap to cover about half the area, and a drop of freshly collected semen is then introduced at the far side, where it is drawn in by capillary action, in very much the same way that tea is drawn into a dunked cookie (Figure 9.2).

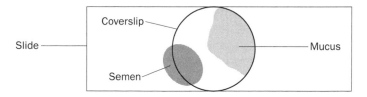

Figure 9.2: Semen and Mucus Being Introduced to Each Other on a Microscope Slide.

I was introduced to this procedure by Dr. Eileen McLaughlin and her staff in the fertility unit at St. Michael's Hospital in

Bristol. I was surprised to see through the microscope that, when the semen and mucus first come into contact, the interface between the two rapidly develops a series of cusps (Figure 9.3).

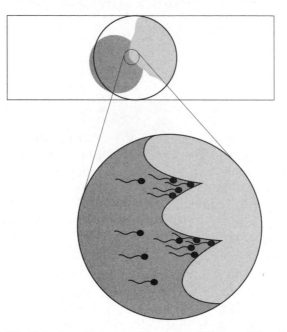

Figure 9.3: The Interface Between Semen and Mucus, Showing Formation of Cusps.

The spermatozoa, I was told, enter the cervical mucus through the tips of these cusps. It was the cusps themselves that intrigued me, though. How were they formed, and why should spermatozoa enter only through the tips?

Although liquid/liquid interfaces are a speciality of mine, I had only seen cusps like this on one previous occasion, when I had dipped an oil droplet into a solution of protein. As more and more protein molecules competed for space at the droplet surface, the sideways pressure that they generated in trying to push each other out of the way was sufficient to collapse the surface into a series of folds. This case seemed different. It appeared that something in the semen was reacting chemically with something in the cervical mucus to form a surface film that spontaneously folded.

The surface film is obviously important, since it could form a barrier to sperm penetration. To find out more about it, I decided to check out what would happen if the semen was replaced by water. To my intense surprise, the same type of cusp appeared. Whatever was going on, it didn't involve chemicals in the semen. My guess was that protein from the mucus was accumulating at the mucus/water interface in the same way that protein had accumulated at the surface of my oil drop. This made some physical sense from the point of view that the tips of the cusps are likely to be the weakest points in such a structure, and hence the most easily penetrated by a wriggling spermatozoon.

My experience of surface chemistry told me that if protein was going to accumulate at the mucus/water interface, it should also accumulate at a mucus/air interface. If this were the case, the rigid protein film should collapse into folds as soon as water touched any part of it. When I looked more closely, I was gratified to find that my scientific instinct had, for once, been right (Figure 9.4).

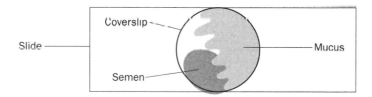

Figure 9.4: Cusps Formed at Air/Water Interface When Semen or Water Comes into Contact with Cervical Mucus.

The folds, which will eventually be covered with water (or semen), are likely to crack at the tips, exposing the underlying mucus. For a spermatozoon to be able to push its way in to this mucus, it needs to exert a pressure greater than the *yield stress* of the mucus, which is (roughly) the pressure at which the mucus gives way. The yield stresses of jelly-like materials are measured in units of pressure called *Pascals (Pa)*. For the purposes of reference, a thin paperback book lying on a desk exerts a pressure of around 100 Pa. The pressure in a car tire is around 200,000 Pa.

The yield stress of cervical mucus at midcycle is around sixty Pascals, a characteristic figure for jelly-like materials that can hold their own shape but are still fairly easy to work (like an industrial hand-cleaning gel, for example). For cervical mucus, Nature has worked things out pretty well. A yield stress of 60 Pa means that any spermatozoon capable of swimming faster than 2 mm/min can push its way through the surface. On either side of midcycle, the yield stress rises rapidly, so that the mucus provides an efficient, though not guaranteed, barrier to conception.

The final word on whether spermatozoa can penetrate the cervical mucus may come from the female partner in a couple, since a link has been discovered between her enjoyment of a particular sexual encounter and the quantity of sperm in the cervical mucus afterwards. This may partly be a matter of orgasm (there is some evidence that spermatozoa are sucked in during orgasm), but seems to embrace many other factors, including the moistness and receptiveness of the vagina. These are, of course, physical factors, so perhaps physics has the last word after all.

Step 3: The Race Is On

Once one spermatozoon has penetrated, others follow. They don't swim off in any old direction, but play follow-the-leader. If several spermatozoa have initially penetrated at different places, the result is lines of spermatozoa traveling in lanes, often at different speeds. The type of mucus in which this happens is evocatively called *motorway mucus*. This is the "best" mucus, with properties that give as many spermatozoa as possible a chance of making it to the uterus after a swim of some thirty millimeters. How, though, can spermatozoa swim at all in this jelly-like material? It would be impossible if the swimming movements of a spermatozoon were reversible (like the oar movements of a rowing boat, or the movements of a swimmer who is swimming "stiff-armed") because any movement that would drive the spermatozoon forward in the glutinous medium would be canceled out by an opposite movement that would drive it right back again (Figure 9.5).

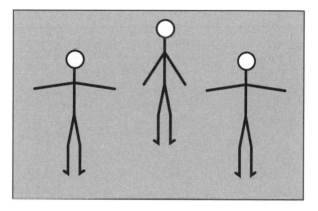

Figure 9.5: The Problem of Swimming in a Viscous Medium Using Reversible Movement of the Arms.
(Left) starting position, *(center)* body moves forward as arms move back; *(right)* body returns to starting position as arms move forward. In a medium of low viscosity, the problem can be overcome in part by making the driving movement faster in one direction than the other, e.g., pulling the arms back rapidly, then moving them forward slowly. If viscosity dominates, this tactic fails.

When I asked Eileen how sperm were thought to manage it, she told me that it was still very much a matter of opinion. Some workers believe that the answer lies in the fine structure of the mucus, which contains long string-like molecules called *mucopolysaccharides*. These workers claim that the mucopolysaccharides form bundles about 0.5 μm in diameter (one-tenth of the diameter of a sperm head), with aqueous channels in between that the sperm can squeeze along. Other authors believe that the mucus itself moves. Some believe that the beating of the sperm tails sets up a resonance with the natural frequency of the mucopolysaccharides, like walkers crossing a bridge in step and setting it swaying, and that the resulting rhythmic waves in the mucus carry the sperm along. Still others claim that the rhythmic beating of the *kinocilia* (tiny hairs that line the cervix) produces waves that enhance sperm migration.

There may be some measure of truth in each of these explanations, but to me none was completely satisfactory, because none took into account how large, string-like molecules actually behave in solution. I had had some experience of this behavior as a food scientist when studying the way that jellies set. Most food jellies are based on gelatin, a long thin molecule that wriggles around freely in hot water. As the water cools, the gelatin molecules begin to form a three-dimensional net, held together by weak links at the points where the molecules randomly cross over each other. If the weak jelly is stirred at this stage, it rapidly becomes more liquid; in other words, it is *shear-thinning*. Stirring disrupts the weak linkages and lines up the long molecules along the flow lines, so they can slide past each other more readily.

I knew that cervical mucus, like edible jelly, is a solution of long molecules that has shear-thinning properties, so that mucus stirred by the lashing tail of the spermatozoon would form a trail of lower viscosity than the surrounding mucus, a line of least resistance for subsequent spermatozoa to follow. This explained the "motorway mucus" effect, but still did not explain the ability of spermatozoa to swim through cervical mucus in the first place. I was puzzling out loud over this question one morning in the departmental coffee room. My new Ph.D student Rachel was busy reading her e-mails, and overheard my question. Later that day she silently handed me a reprint of a 1976 article that told me everything I wanted to know.

The article was the transcript of a talk by E. M. Purcell called "Life at Low Reynolds Numbers" — in other words, life under circumstances where viscous drag dominates every attempt to move. Organisms like spermatozoa manage to swim under such circumstances by using the tail (or *flagellum*) as a flexible oar, or whip. This dodges the problem that a human swimmer with rigid arms or a rower with rigid oars would experience, because a flexible arm or oar can bend one way during the first half of a stroke and then change shape to bend the other way during the second half.

Some small organisms, such as the biflagellate alga *Chlamydomonas,* use their flexible flagellae in this way to do the breaststroke (Figure 9.6). Human sperm do it in a different way, lashing their single flagellum from side to side under the drive of a molecular motor. The tail lashing does the trick be-

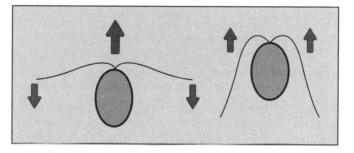

Figure 9.6: Swimming Motion of *Chlamydomonas*.

cause it flexes so that a two-dimensional wave appears to be continually traveling along it. Even though viscous drag always opposes the direction of motion, Purcell showed that a tail moving in this way can, incredibly, use viscous drag to propel itself forward. A fuller explanation is given in Figure 9.7. The diagram may look complicated but, like many scientific diagrams, it is relatively simple to follow if taken step by step.

It takes 10–15 minutes for the fastest human sperm to swim the length of the cervical canal. A few spermatozoa make it much more quickly by a process called *rapid transport,* which can carry inert particles through within two minutes. No one knows how this process works, but the sperm that take advantage of it are not as lucky as they appear, because the time

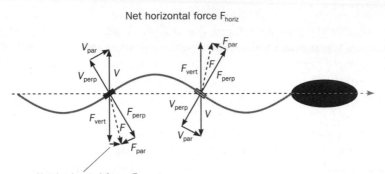

Figure 9.7: Viscous Drag on a Flagellum.
The flagellum is the wavy line. The segment portion shown as a filled rectangular box is moving upward with velocity V; the open box portion is moving downward with velocity V. Each velocity can be broken into two separate velocities — one parallel to the segment, and one perpendicular to it. The principle is the same as that of the knight's move in chess, which is a steep diagonal move that can also be thought of as two steps parallel to one side of the board, followed by one step parallel to the other side. For the filled rectangular segment, the two separate velocities are marked V_{par} and V_{perp} (with similar nomenclature for the clear rectangular segment). The forces F of the viscous drags that oppose these velocities are labeled F_{par} and F_{perp} respectively.
    Now for the clever bit. We recombine F_{par} and F_{perp} to give an overall drag force F, and then break this down again, but this time into a force parallel to the horizontal axis (F_{horiz}) and one perpendicular to the axis (F_{vert}). The perpendicular one acts downwards, and the parallel one acts forwards. When we go through the same procedure for the open segment, the final perpendicular component F_{vert} of the viscous drag acts upward, canceling the effect of that on the filled segment. The parallel component F_{horiz}, however, acts forward, adding to the parallel component acting on the filled segment. The overall viscous drag, then, acts to push the spermatozoon forward.

that their companions spend in the cervix is needed for chemical changes that allow them to penetrate and fertilize the egg.
    Those spermatozoa that make it to the uterus have yet more problems to overcome. The first is similar to that experienced by surfers on some Sydney beaches — the water is full of sharks. The "sharks" are white cells produced by the immune system, and whose job it is to scavenge foreign cells and other materials not recognized as "self." Nothing could be more foreign than a spermatozoon in a uterus, and lucky indeed is the spermatozoon that makes it to the neck of one or other Fallopian tube. Of the original few hundred million, only a few

hundred are now left. Most of these have made it through the uterus by surfing on uterine waves, induced by prostaglandins in the seminal plasma. The waves seem to go towards the oviduct *and* the cervix, so the spermatozoon has to catch the right one. Once in the Fallopian tube, the spermatozoa begin jostling for position like adolescent youths at a street corner, waiting for the egg to arrive. When it does, it is plastered in makeup — a layer of jelly so thick that it takes all of the efforts of the spermatozoa to penetrate it (Figure 9.8).

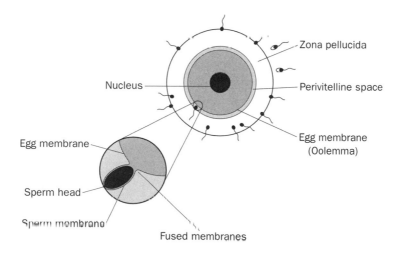

Nucleus
Zona pellucida
Perivitelline space
Egg membrane
Egg membrane (Oolemma)
Sperm head
Sperm membrane
Fused membranes

Figure 9.8. Fertilization of the Egg.
This is a very simplified picture, giving the main outlines only. Even the picture of the egg itself is simplified. In real life, it is surrounded by a cloud of cells that includes *cumulus cells*, which secrete a matrix of hyaluronic acid, similar to the lubricating material in the joints.

The process is a complicated one, involving enzyme action and biochemical and structural changes. In outline, the sperm first has to drill through the surrounding cell layer (called the *cumulus oophorus*) using a combination of mechanical movement and chemical dissolution. Once through, the sperm encounters a rigid layer, the *zona pellucida*, where the cap of the sperm that was the drill bit is lost in a complicated process called the *acrosome reaction*. The sperm now has to work through the *zona* in its more-or-less naked condition. It does this by using its asymmetric tail motion to rock the blade-shaped head back and forth, generating sufficient force to break individual molecular bonds as the head levers and cuts its way through.

The sperm eventually reaches a gap called the *perivitelline space,* where the thin membrane surrounding it makes contact with the membrane surrounding the egg (*oolemma*). The two membranes fuse, and the whole sperm, with its DNA-carrying head, is engulfed by the egg. The winning spermatozoon is like a knight of old, scaling the defended ramparts and eventually breaking through to the maiden within. Like any cautious knight, it takes the precaution of turning the key on the inside of the door as soon as it is through, by releasing a burst of calcium that irreversibly alters the egg membrane so that other sperm cannot enter while the winner is combining its DNA with that of the egg nucleus.

A new life has now begun. Perhaps it is no more than Darwinian evolution at work which dictates that the sperm with the greatest mastery of physics is the one which has the best chance of initiating that life.

coda

I have attempted to convey something of what it feels like to be a scientist, with the daily task of working out how some small part of the world functions. We enjoy ourselves, just as anyone in a satisfying and rewarding occupation does, but underneath the pleasure and excitement there is a real sense of purpose. As the stories in this book have shown, there is always the possibility that an answer to even the most trivial-sounding question might help to produce a new insight into the nature of the world in which we live. Scientists live for these moments.

People outside science often picture scientists as solitary geniuses, occupying pinnacles far above the rest of us. That may be true for a few, just as it is true for a few musicians or artists or writers. Fortunately for most working scientists, including myself, it is perfectly possible to make useful contributions to science without being a genius. This is because science is largely a communal activity, to which people with many different skills contribute. These include some who are good with their hands, compulsive gatherers and arrangers of facts, persistent seekers of answers to niggling questions, those with a "feel" for animals or rocks or plants, and many others. All have their parts to play. Our main common characteristic is an openness to criticism, albeit sometimes with gritted teeth. We share results and ideas through open publication, which sometimes leads to more criticism, but produces an awareness of important questions, and acts as a resource of information that has passed the critical test and which can lead to fresh insights. Without this communal sharing, there would be no such resource, and ultimately no science.

Would it matter if there were no science? It would to me and my fellow scientists, of course, because science provides our livelihood. It would also have mattered to me personally because I would be dead by now but for the invention of modern antibiotics. This and other practical benefits for our lives and lifestyles are often advanced as major reasons for supporting and investing in science, and it cannot be denied that we are immeasurably better off, in this sense, as a result of scientific discoveries. The practical results of science extend well beyond the biomedical and communication fields that tend to dominate the headlines. Even the shelves in our supermarkets would be relatively empty if scientific advances had not let us understand and control decay processes in plant and animal products.

Practical benefits, though, are outweighed in many people's minds by practical problems. Of what use is a longer life, or even a loaded supermarket shelf, when that life and lifestyle are threatened by an atomic bomb or a genetically engineered virus?

Some people, when faced with such problems, take the extreme view that this science should have been stopped earlier, and should be stopped now, before we get into even deeper waters. Others believe that man's best (and possibly only) chance is to press on, seeking further understanding of our world ourselves and the ways in which we might use science for good rather than ill.

A compromise, supported by the majority of people, is to encourage the pursuit of science, but to direct scientific research towards peaceful applications only. Given that it is impossible to quell our natural human curiosity, they ask, surely it must at least be possible to direct that curiosity towards peaceful ends? Unfortunately for those of us who wish only for peace and cooperation, the answer is "no." Even if all direct military research could be stopped, we would be little further forward, because it is simply not possible to predict where an advance in scientific understanding might lead. A prime example occurred in 1939, just before the outbreak of the Second World War, when two German scientists published the key theory which made the atom bomb possible, in the British journal *Nature*. Only when American and British scientists began to consider whether an atom bomb could be built did this finding assume significance. If its consequences had been obvious, the German group would not have revealed their discovery so openly.

The consequences of any particular scientific discovery are often not obvious, even to the discoverer. Rutherford, commenting on the possiblility of splitting the atom (the first step along the road to the atom bomb), was quoted as saying, "Anyone who thinks that any practical benefit can be gained from splitting the atom is talking moonshine." Nor was he looking for practical benefit. The prime reason for attempting to split the atom was to find out what was inside, and thence to understand a little more about the material from which we and the whole universe are built. The technological consequences (so far) have included, not only the atomic bomb, but nuclear power stations (whose utility some will debate), new methods of revealing and healing diseased and abnormal tissue, and a truer picture of the nature and origin of the universe. In the future they may include "clean" atomic power through nuclear fusion, and even ways of reaching distant stars and saving the human race from extinction as the Sun's own nuclear power plants run down. Who is to say, then, whether the consequences of Rutherford's discovery will ultimately be "good" or "bad"?

I myself believe that curiosity-driven scientific research is not intrinsically "good" or "bad." The motivations of individual scientists may sometimes be classified in this way, but in general it is so difficult to predict the answer to any seriously asked question, let alone any applications of that answer, that ethical judgments on the question or questioner are simply inappropriate. Such judgments only come in when we begin to think of how we might use the answers, and they involve not only scientists, but the whole community. It is the applications that we need to control, not the process of discovery.

This is *not* to say that scientists should abdicate responsibility, but the responsibility should be placed squarely where it belongs — in sharing their knowledge with the wider community, so that the community and its political leaders can make informed judgments. Scientists can do no more than that, but they should do no less.

One of the prime things that we need to share, and which I hope that this book has gone some way towards sharing, is how science actually works. The fact that we cannot predict which questions will produce trivial answers (or no answers at all) and which questions will produce significant answers means that we need people with many different approaches to provide the variety from which a few major results will emerge. The problem is akin to that of genetic diversity in the wild. If we end up with a scientific monoculture, with everyone working on the same few questions (those perceived to be important by politicians, businessmen, or technocrats), the result will be no science at all.

Unfortunately, that is the way that things seem to be heading. The direction of science is largely in the control of committees who dole out money for specific projects. Pressure from the money providers (mostly government and industry) increasingly means that support is not forthcoming unless the applicant can show a reasonable prospect of getting a "useful" result. The inevitable consequence is a focus on problems where the answer is either known or is predictable with such assurance that it is hardly worth checking. Genuinely important questions are pushed to the margins, eventually to become extinct, and, consequently, the diversity of science decreases daily.

The above process might have some merit if the answers to particular scientific questions could lead to predictable advances in technological application. As many of the examples in this book have shown, though, the most important applications of scientific advances are seldom related to the original scientific motivation. The American biochemist Hans Kornberg once drew up a list of the ten most significant medical advances of the twentieth century. Seven out of the ten arose from research that had nothing to do with the eventual application.

The driving force for genuine research comes from each individual scientist's perception that a question is *important,* and worth investigating. Many questions turn out to be unimportant, but we cannot always tell in advance which is which. The very best that we can do is to encourage scientific diversity. All attempts to pick out in advance which questions are worth pursuing, and to focus only on those, will produce a lesser result.

With diversity comes responsibility, which is not that of the scientist alone, but also that of the community of which the scientist is a part. At the moment, scientists are too often apart rather than a part, and other members of the community can consequently feel suspicious and disempowered. It is up to us, as scientists, to share what science is about, and what it can and cannot do, with the rest of our community. I hope that this book has at least taken one small step along the way.

appendix 1:
mayer, joule, and the concept of energy

Mayer's unlikely inspiration for the concept of "energy" was the sight of a horse sweating as it pulled a load up a hill. His key idea was that the horse was getting hot, not because it was moving, but because of the physical work that it had to do to generate the movement. He thus turned the question of the relationship between *heat* and *movement* into a question of the relationship between *heat* and the *physical work* needed to produce movement. Having reframed the question, he drew three far-reaching conclusions. The first was that when we use heat to do work (e.g., to drive a steam engine) or perform physical work to generate heat (e.g., when we rub our hands together), the two are literally transformed into each other. Heat becomes work, and works becomes heat. That idea was remarkable enough, but he went further to deduce that heat and work must be inter-convertible in a constant ratio. Otherwise, he argued, we could use an initial small amount of heat to do as much work as we want, simply by using the work produced initially to generate more heat than we started with, and so on. Such processes don't work, or else we could fly a Boeing 747 by striking a match. There is no such thing as a free lunch.

Mayer tried to prove his thesis by arranging an experiment at a paper factory where the pulp in a large cauldron was stirred by a horse going around and around in a circle. Measuring the rise in the pulp temperature, he obtained a figure for the amount of heat produced by a given amount of mechanical work done by the horse.

Mayer's experiments were crude, and he does not receive any credit for them in modern textbooks, despite having been the first to conceive or attempt them. The credit goes instead to an English brewer named James Prescott Joule, who used a paddle wheel driven by a falling weight to stir a bucket of water, finding that the temperature of the water rose by twice as much when the weight dropped twice as far, so that twice as much physical work was generating twice as much heat.

Joule deserves his credit, since he was the first to perform experiments whose results were sufficiently accurate and reproducible to be convincing. It is more difficult, though, to reject Mayer's claim to priority for his third and most important insight, which was that heat and physical work are not only transformable into each other, but are actually different forms of the same thing. We now call that thing energy.

What Mayer had enunciated became what we know now as the principle of *conservation of energy*, the cornerstone of modern science. Yet his name is hardly ever associated with it. Mayer committed the cardinal sin — he was an outsider. Some establishment scientists of the time defended his priority, but others, especially the British scientist Peter Guthrie Tait, poured xenophobic scorn on his rather metaphysical style of argument as "subversive of the method of experimental science." Even his fellow German, Hermann Helmholtz, initially a defender of Mayer's innovativeness, subseqently derided his "pseudo-proof." Under these criticisms, Mayer attempted suicide by jumping out of a window thirty feet above the street. Luckily, his lack of experimental ability was once again proven and he survived the attempt. He was eventually honored as a genuine innovator, although his methods lacked the rigor needed to convince others of the correctness of his remarkable insights.

Mayer was a man of ideas. Joule was a man of action, eager to confront ideas with hard facts. If Joule's experiments to prove that the heat generated by physical work is in direct proportion to the amount of work done were truly correct, then the rest of Mayer's logic follows.

First, though, Joule had to decide what "physical work" meant. Mayer had been quite vague about it. Obviously, "physical work" is something that we do when we move an object by pushing or pulling — in other words, by applying a force. Joule decided, apparently intuitively (and following Carnot's earlier ideas, published in 1824 under the title "Reflections on the Motive Power of Fire"), that the amount of work that we do in moving an object depends only on the force that we apply and on how far the object is moved. The further we push, and the harder we have to push, the more work we are doing. In simple mathematical terms: work = force × distance. The definition makes good intuitive sense. When we have to push a car whose engine has failed, for example, the work that we feel we are doing surely depends on how hard we have to push and how far we have to push the car. The further we push, or the harder we push, the more work we feel we are doing.

The intuitive idea of "work" underlies almost the whole of modern physics. All of our measurements of "energy," for example, rely eventually on measuring how much work the energy can be made to do. Since energy is now believed to be the stuff of the universe, the correct definition of "work" is crucial. Joule's intuitive guess at the definition, apparently so simple, has turned out to be the one that provides a totally self-consistent picture, and is the one that we still use. It is remarkable that the whole of modern physics, rigorous as it is, has as its foundation a totally intuitive guess.

Joule was able to demonstrate his guess practically by constructing his famous paddle-wheel experiment. Shortly after he had completed it, Joule married, and took his new bride for a honeymoon at the famous Chamonix falls in the Swiss Alps. One can imagine his wife's chagrin when she found that Joule had secreted a thermometer in the baggage, intending to measure the temperature of the falls as the water fell through different distances. It can only have been matched by Joule's chagrin when he found that any increase in temperature was offset by the effects of the cold air as it carried the heat away.

Mayer's insight has now been extended to encompass the idea that all forms of energy can be transformed into each other. All of them, for example, can be transformed into heat and hence used in cooking. The final step was made by Einstein when he showed that matter itself may be regarded as compressed energy, transformable directly into the heat of a nuclear power station or the heat and light of an atomic bomb. Mayer's logic is formalized these days into the first of the three laws of thermodynamics. This law (well known, even if its meaning is not always fully appreciated) states that energy cannot be either created or destroyed. The second law states that converting any form of energy to work is never one hundred percent efficient (some is always converted to heat), except at a temperature of absolute zero. The third law states that we can never reach absolute zero. These laws are equivalent to the laws of gambling in the old Wild West:

1. You can't win.
2. You can't break even.
3. You may not leave the game.

appendix 2:
the effect of temperature on food molecules

The effect of temperature on food molecules depends on the type of molecule. There are four main types of molecule to consider — water, fats, carbohydrates, and proteins. From the point of view of the chef, water was an unfortunate choice by Nature as the universal, dietetically innocuous liquid base for our existence. Essential for life it may be, but the gastronomically unfortunate fact is that it is not only boring in its contribution to flavor, but also boils at a temperature where hardly any of the interesting processes in cooking occur. The convenience of the temperature is such, nevertheless, that boiling water is frequently used in cooking as a safe and efficient method of transferring heat. Only the surface of the food initially reaches the boiling temperature of 100°C. The secret of using boiling water is to time the cooking so that the inside of the food only heats up to the ideal cooked temperature (usually well below 100°C).

Water in the food itself also frequently contributes to other desirable changes. When we cook rice, pasta, or potatoes, for example, water acts not just as a medium to transfer heat but also to help change the very texture and palatability of the cooked food.

If higher temperatures are desired, fats and oils can be used as heat transfer media. They also occur as important dietary components of many foods. Most of the flavors that we perceive are carried by the oily component of the food (this is why diet foods, lacking oil, can taste so bland), and the oil when digested releases twice as much energy as an equivalent quantity of carbohydrate.

An oil is simply a fat in the melted, liquid state. There is a big difference, though, between the oils found in foods and those

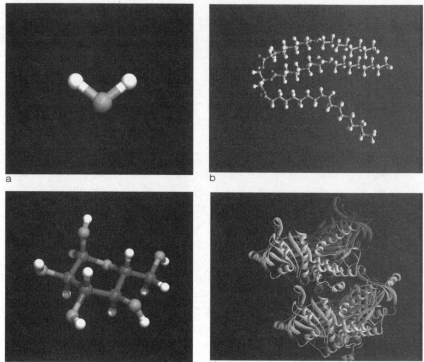

a

b

c

d

Figure A.1: Computer-Generated Pictures of Different Types of Food Molecule. A Molecule Consists of Atoms (Represented Here as Spherical Balls) Linked Together by Chemical Bonds (Represented Here by Sticks Joining the Atoms).

These diagrams are intended to show the shapes and relative complexities of the molecules that go to make up our food.

a. *Water* — Here there are just three atoms — an oxygen atom in the center, linked to two hydrogen atoms.

b. *Fat* — Fats contain three long hydrocarbon chains (i.e., chains of linked carbon atoms with hydrogen atoms attached to each carbon atom), with all three chains attached to a common glycerol head-group (on the left of the diagram). They typically contain several hundred atoms. The particular fat shown is an unsaturated fat, which means that a bond in at least one of the chains is doubled up. This has the effect that the chain itself doubles up, as if kicked in the stomach. The one shown here has such double bonds in two of its chains.

c. *Carbohydrate* — Carbohydrates are built from small sugar molecules such as glucose, which comprises a ring of five carbon atoms and one oxygen atom, with other oxygen and hydrogen atoms attached to its periphery. It contains 24 atoms. Most carbohydrates (such as the starch in potatoes and cereals) consist of many such rings joined in a line, and can contain tens of thousands of atoms.

d. *Protein* — Proteins, such as the egg albumin shown here, can also contain thousands of atoms, joined together in chains that can fold to form helices, flat sheets, or an apparently random mess. The mess can be resolved by concentrating on the shapes of the chains rather than the positions of the individual atoms. Here are four albumin molecules with the shapes they adopt in space revealed by this technique.

used to lubricate car engines. It is the difference between Jake the Peg and Long John Silver. Food oil molecules, like Jake the Peg, have three legs attached to a common backbone. Lubricating oils, like Long John Silver, have only one. The shape of the legs affects the thermal behavior of the oil or fat. Each "leg" consists of a jointed chain of carbon atoms. At sufficiently low temperatures, the legs can pack side by side to form a solid crystal. Fat molecules with straight legs ("saturated" fats) pack closely, vibrate little, and require a relatively high energy to separate them. Those with bent legs ("unsaturated" fats) pack less well, are easier to separate, and melt at a lower temperature. Some, like those from peanut and safflower, may even be oils at room temperature.

Fat and water molecules are relatively small, and the effect of temperature on their culinary behavior is relatively simple, consisting largely of freezing or melting. Protein and carbohydrate molecules are usually much larger. A typical fat molecule may contain a couple of hundred atoms, whereas proteins and carbohydrates may contain many thousands. The smallest carbohydrate molecule, glucose, is actually rather smaller than a fat molecule. Plants store glucose, however, not as single molecules, but mostly as long chains of linked molecules in the form of starch. These chains in turn pack into elaborate structures, where crystalline layers alternate with amorphous layers in a series of concentric rings, beautiful to observe under the microscope.

All of this natural beauty is undone when we cook starchy foods such as rice, pasta, or potatoes. As the temperature increases, water molecules gradually insinuate themselves between the starch chains. The delicate energetic balance between the attraction of the chains for water or for each other is gradually shifted until, at a particular temperature (62°C for potato starch), the chains suddenly separate hugely to accommodate an influx of water that turns the starch granules from hard gritty particles to a soft swollen jelly. Higher temperatures don't produce any further effect — potatoes, for example, can be cooked quite comfortably at 70°C. It just takes longer for the heat to reach the center.

Meat and fish are major sources of proteins, which generally require lower temperatures than carbohydrates for the disruption of their structure. The energy needed to disrupt the three-dimensional structure of many folded protein chains, for example, corresponds to a temperature of around 40°C (this is why our body temperatures are locked at 37°C). A high-quality steak, such as fillet or sirloin, consists principally of the muscle proteins actin and myosin, which adopt an extended configuration in the relaxed muscle that is maintained by weak cross-links. The muscle proteins shorten when their cross-links are broken; a process that makes the meat tougher. Such steaks should not be heated at the center to temperatures much above 40°C, which is why we use such short cooking times for them.

The proteins in connective tissue (the white strands and sheets in meat) require much higher temperatures for disruption of their three-dimensional structure. The principal component of connective tissue is the protein collagen, which actually consists of three protein chains, wound around each other to give a strong rope-like structure, which is why meat with a lot of connective tissue in it is so tough. To disrupt this structure requires a temperature of around 60°C, together with the presence of water. A combination of the two converts the triple-stranded collagen into single-stranded gelatin, which is much softer and more digestible. Unfortunately, at such a high temperature the muscle proteins will have become very tough. The solution, such as it is, is either to choose high-quality meats or to accept a compromise in texture.

At even higher temperatures, the kinetic energy of the molecules can become greater than the energy of the bonds between individual atoms in the chain. The chain may break up, and the bonds can become available to form new links with other molecules. This is what happens in *browning*, where the kinetic energy at temperatures above 140°C is so high that protein and carbohydrate chains can form complicated new cross-links with each other, producing new materials that are both brown and tasty. We start our roasts off at high oven temperatures to promote these browning reactions.

notes and references

chapter 1: the art and science of dunking

Page
1 *Market research on cookie dunking.* Informal survey conducted in London in connection with the "cookie dunking" project.
2 *. . . an elastic net of the protein gluten.* Network was defined tongue-in-cheek in Johnson's 1755 *Dictionary of the English Language* as "anything reticulated or decussated at equal intervals, with interstices between the intersections." It is the interstices, or cross-links, that count; without them there would be no net. For the last fifty years it has been believed that these cross-links in wheat-flour dough are formed by disulphide bonds. Only recently has it been discovered (Tilley, K., *Journal of Agricultural and Food Chemistry*, vol. 49, p. 2627) that they are formed by cross-links between the totally different tyrosine residues, and that the disulphide idea was not based on hard evidence. It just goes to show that "folk explanations" are as likely to penetrate science as anywhere else. The difference is that, in science, there is at least a way of eliminating them.
3 *. . . seven thousand calls in a quarter of an hour.* Not all of the calls were about dunking. Many inquirers wanted to know about the science underlying other familiar activities, such as cooking, sun bathing, and using mobile telephones. The supposedly abstruse and difficult scientific principles involved in the answers seemed to pose little difficulty for the questioners, since the principles were being related to something with which they were already familiar.
3 *Robert Hooke and the Royal Society.* "Hooke, Robert," article in *Encyclopaedia Britannica*, 11th edition, Cambridge University Press, Cambridge, 1911.

   The exact terms of Hooke's appointment, as recorded in the Journal Book of the Royal Society for 5 November 1662, were: "Sir Robert Moray proposed a person [Hooke] willing to be employed as a curator by the Society, and offering to furnish them every day, on which they met, with three or four considerable experiments, and expecting no recompence till the Society should get a stock enabling them to give it."

"Considerable," it transpired, meant "original." And today's Ph.D students think that they have it tough!

4 *The structure of DNA.* Watson, James D., *The Double Helix,* Atheneum, New York, 1968.

5 *Francis Bacon and the nature of scientific research.* Bacon, Francis, *The Novum Organon,* or, *A True Guide to the Interpretation of Nature,* Kitchen, G. W. (transl.), Oxford University Press, Oxford, 1855. An amazing number of people still think that science works in the way that Bacon suggested.

5 *Paradigm shifts in science.* Kuhn, Thomas, *The Structure of Scientific Revolutions,* 2nd edition, Chicago University Press, Chicago, 1970.

5 *Proof in science.* Medawar, Peter, *The Limits of Science,* Oxford University Press, Oxford, 1985.

6 *The mathematics of a drunkard's walk.* "Drunkard's Walk Helps Unfold Secret of Polymers," *New Scientist* magazine (136), 12 December 1992. This article summarizes the principles of random walks in an accessible way, and shows how the square relationship and the factor of four that follows arises. There are also numerous Web sites that cover the subject of random walks at different levels of sophistication.

7 *Washburn's experiments (and equation).* Washburn, E. W., *Physics Review* (17), 1921, p. 374.

8 *Graphs, symbols, and equations.* These are an immense source of confusion to non-scientists (and to some scientists!), but they needn't be. The principle is very simple. An *equation* describes how one thing depends on another. If a car is traveling at 60 kilometers per hour, for example, then the equation (distance in kilometers) = 60 × (number of hours) shows just how the distance traveled depends on the time spent.

To avoid writing everything out in full, scientists use abbreviations. D is a common abbreviation for distance, and t is the universal abbreviation for time. The equation above thus becomes $D = 60 \times t$, which is much easier to write and just as easy to read with a little practice.

A *graph* is simply another way of writing an equation to display visually how one thing depends on another. By convention, the thing that is depended on (in this case, the time) is drawn along the bottom (horizontal) axis, and the thing that depends on it (the distance) occupies the vertical axis. It's that simple.

10 *Young's version of the Young-Laplace equation.* Young, Thomas, *Philosophical Transactions of the Royal Society of London* (95), 1805, p. 65. See also Young, Thomas, *A Course of Lectures on Natural Philosophy and the Mechanical Arts* (2 vols.), J. Johnson, London, 1807.

Young hated mathematical symbols and wrote his equations out entirely in words, which makes reading his papers incredibly hard going.

11 *Laplace's version of the Young-Laplace equation.* Laplace, Pierre Simon de (Marquis), *Supplément au dixième livre du traité de mécanique céleste,* 1806. Translated and annotated by Bowditch, N. (4 vols.), 1829–1839, Boston. Reprinted by Chelsea Publishing Co., New York, 1966.

12 *Scientists investigating familiar phenomena.* This has frequently led to important and fundamental discoveries. The famous story that Newton discovered the universal law of gravitation after being hit on the head by a falling apple unfortunately has no foundation (although it is interesting to note that the modern unit of force [the Newton] is approximately equal to the force of gravity on an average-sized apple). There are, though, plenty of real examples of universal laws being derived from the observation of commonplace phenomena. These include: Galileo's discovery of the pendulum laws after observing the swinging of a chandelier in the cathedral of Pisa; Mendel's deduction of the laws of genetics from his observations of peas growing in a garden; Rumford's hypothesis that heat is a form of motion, a conclusion that he came to after observing the enormous amounts of heat generated during the boring of brass cannons (anyone who has ever had occasion to drill a hole in a piece of metal will be aware of this phenomenon on a smaller scale). In modern times, the universal theory of chaos, which dominates such diverse topics as the growth and decline of animal populations and financial movements in the stock market, originated from Lorenz's efforts to understand weather patterns.

12 *Mean radius of curvature.* I have used the technical term "mean" here so that my fellow specialists in the field don't shoot me down. Most menisci are curved differently in different directions, and the calculation of a "mean" is used to account for this fact.

14 *Poiseuille's equation.* The equation is simply:

$$L^2 = \frac{\Delta P \times R^2 \times t}{4\eta}$$

where L is the distance traveled by a liquid of viscosity η in time t along a cylindrical tube of radius R under a pressure head ΔP (Poiseuille, J. L. M., *Comptes Rendus de l'Académie de Sciences, Paris* (11) 961, 1041 (1840); (15) 1167 (1844)). Note: Δ is the usual scientific shorthand for "a change in."

16 *Experiments on the swelling of individual starch granules.* These were reported in Fisher, L. R., Carrington, S. P., and Odell, J. A., "Deformation Mechanics of Individual Swollen Starch Granules," *Starch, Structure and Functionality* (P. J. Frazier, P. Richmond, and A. M. Donald, eds.), Royal Society of Chemistry (London), Special Publication, no. 205, 1997, p. 105.

18 *Stress.* The term has a precise technical definition, which in this case is just the force divided by the area over which it is applied. Since the area at the crack tip is tiny, the stress is huge.

18 *The science of how cracks form and grow.* The phenomenon is discussed in an entertaining and simple fashion in *The New Science of Strong Materials,* Gordon, J. E., 2nd edition, Pelican, London, 1976, which also gives a photograph of the *Majestic's* near-disaster. A picture of the *Schenectady* actual disaster is given in *Structures,* Gordon, J. E., Pelican, London, 1978.

20 *Media coverage of cookie dunking.* The story was featured in all major British newspapers over 24–25 November 1998, appeared on TV and radio news worldwide, and even found a place in the *Wall Street Journal.* It also became the subject of numerous features.

21 Nature *article on cookie dunking.* Fisher, Len, "Physics Takes the Biscuit," *Nature* (397), 469, 1999.

chapter 2: how does a scientist boil an egg?

Page

23 *James Bond's gourmet pretensions.* These are amusingly dealt with by Kingsley Amis in *The James Bond Dossier,* Jonathan Cape, 1965.

24 *Nicholas Kurti.* His remarkable life is documented in *Biographical Memoirs of the Royal Society of London* (46), 2000, pp. 299–315.

24 *The Physicist in the Kitchen.* This was the topic of one of the famous Friday Evening Discourses presented at London's Royal Institution (founded by Rumford). Nicholas broke with tradition in having his lecture televised live, and also by refusing to be locked up beforehand, a tradition that developed after a lecturer in the last century (Charles Wheatstone) took fright and ran away before the lecture.

The actual date of the broadcast was Friday, 14 March 1969. Unfortunately, the BBC has destroyed the only tape of this historic event. The only record is the script, published in *Proceedings of the Royal Institution* (42), no. 199.

26 *. . . the sensation of heat is caused by particles of caloric passing into our bodies.* Maunder, S., *Scientific and Literary Treasury,* 1841.

26 *The life of Benjamin Thompson.* This has been written up in many places. A particularly interesting account is given in the ineffable 11th edition of the *Encyclopaedia Britannica* (article on "Rumford, Benjamin Thompson, Count," *Encyclopaedia Britannica*, 11th edition, Cambridge University Press, Cambridge, 1911). When Rumford gained his title, his American origins asserted themselves, and he took the title of "Count Rumford" in recognition of his wife's home town of Rumford, New Hampshire, now known as Concord and the state capital.

Rumford was an insatiable observer of life's minutiae, and a prime example of a scientist who used observations of commonplace phenomena as a basis for advancing our scientific understanding. His personal philosophy, very appropriate to the present book, was that: ". . . in the ordinary affairs and occupations of life, opportunities [often] present themselves of contemplating some of the most curious operations of Nature . . . a habit of keeping the eyes open to everything that is going on in the ordinary course of the business of life has oftener led . . . to useful doubts, and sensible schemes of investigation and improvement, than all the more intense meditations of philosophers . . ."

One of the "ordinary affairs" in which Rumford interested himself was cooking. He wished "to inspire cooks with a just idea of the importance of their art, and of the intimate connection there is between the various processes in which they are daily concerned, and many of the most beautiful discoveries that have been made by experimental philosophers in the present age."

The story of Rumford and the bread oven was brought to my attention by the food writer Harold McGee, who reported it in his fascinating book *The Curious Cook*, North Point Press, New York, 1999, p. 22.

26 *The history of the demise of caloric.* This is fascinating in its own right as an example of how science really works — not by definitive experiment and immediate acceptance of a new idea, but by test and countertest, argument and counterargument, and above all by openness, a too frequent casualty in today's world of industrial and military secrecy. The history is discussed in most standard histories of science, and, for those with a scientific background, Harman, P. M., *Energy, Force and Matter,* Cambridge University Press, Cambridge, 1982, and Brush, S. G., *The Kind of Motion We Call Heat,* North-Holland, 1976.

26 *. . . it was considered prudent that he should seek an early opportunity of leaving.* See "Rumford, Benjamin Thompson, Count," *Encyclopae-*

dia Britannica, 11th edition, Cambridge University Press, Cambridge, 1911.

27 *Rumford's observations on boring cannons.* These are reported in his *Collected Works,* vol. II, essay IX. Read before Royal Society, 25 January 1798.

28 *Mayer's ideas.* These were first reported in *The Motions of Organisms and their Relation to Metabolism,* published in 1824 (reprinted in Lindsay, R. Bruce, *Energy: Historical Development of the Concept,* Dowden, Hutchinson Ross, Inc., 1975).

28 *The notion of heat as motion.* This was finally codified by John Tyndall in *Heat: A Mode of Motion,* 6th edition, Longmans, Green & Co., London, 1880, a book in which he also presented a spirited defense of Mayer's contribution.

29 *Einstein on heat and temperature.* The quote is from Albert Einstein and Leopold Infeld in *The Evolution of Physics,* Cambridge University Press, Cambridge, 1938. The presence of Infeld as a coauthor provides another insight into how science really works. When Einstein moved to Princeton, he made it a condition of his employment that a second position be created so that he would have someone to talk to. The person who gained that position was Infeld. The story shows that even scientists such as Einstein do not live in ivory towers. Communication and exchange of ideas is the name of the game, for Einstein and for almost every scientist I have known.

29 *Temperature as average kinetic energy.* To be precise, the energy of a molecule is given by $k \times T$, where k is a number called Boltzmann's constant. The multiple weak intrachain links that maintain the three-dimensional structures of long carbohydrate and protein molecules typically require an energy of a few kT to break them. To break a chemical bond within the chain requires an energy of around 80 kT.

31 *Rumford's discovery of convection.* This is described in Rumford's *Collected Works,* vol. II, essay VII. Quoted in Magie, W. F., *A Source Book in Physics,* Harvard University Press, Cambridge, Mass., 1965, p. 146.

33 *Chefs and the rules of conduction.* I am again indebted to Harold McGee for the story of an informal poll by the American food writer Edward Behr which resulted in a consensus among the chefs polled that a fish fillet twice as thick as another would take less than twice the time to cook (see *The Curious Cook,* North Point Press, New York, 1999, p. 33). Harold has recently produced a computer model of the temperature distribution in a piece of steak cooked under various conditions (McGee, H., McInerney, J., and Harrus, A., *Physics Today,* November 1999, p. 30).

34 *Solutions to Fourier equation for objects of different shape.* These are discussed in an understandable manner by my colleague and fre-
quent coauthor on food matters, Dr. Peter Barham, in his book
The Science of Cooking, Springer-Verlag, 2000, p. 43.

35 *The "interesting mathematical reason."* In this instance the reason for
the range of agreement between the square rule and classical
rules for cooking times comes from the fact that, if the cooking
time t is proportional to the square of the thickness d for a slab of
meat, then a small increase Δd in the thickness means that the
cooking time will increase by a factor $(d + \Delta d)^2 / d^2$, which is equal
to $(d^2 + 2d + (\Delta d)^2) / d^2$. The point, as those who have learned calculus will know, is that if $(\Delta d)^2$ is very small compared to the
other two terms, and can be neglected, then the cooking time increases linearly with thickness, just as predicted by Mrs. Beeton
(and my mother). The point is made in a different way by P. B.
Fellgett in Kurti, N. & G. (eds.), *But the Crackling Is Superb,* Adam
Hilger, 1988, p. 40 (an anthology on food and drink by Fellows
and Foreign Members of the Royal Society).

37 *Richard Gardner.* His temperature measurements on boiling eggs
are reported in Kurti, N. & G. (eds.), ibid., p. 53.

39 *Charles Williams.* His calculations of egg-boiling times, and Hervé
This's comments on the relative setting temperatures of white and
yolk, were reported in *New Scientist,* 13 June 1998.

40 *The effects of temperature on food molecules.* For more detail see, for
example, McGee, H., *On Food and Cooking,* Fireside Books, New
York, 1997, and, Barham, P. J., *The Science of Cooking,* Springer-
Verlag, 2000.

chapter 3: the tao of tools

Page

43 *Woodworking Tools and How to Use Them.* This classic book was re-
worked by Jack Hill (David & Charles, Newton Abbot, 1983).

44 *The Greek author called the "pseudo-Aristotle."* This author derived
the law of the lever around a hundred years before Archimedes,
but the derivation only stands up to critical examination if it is as-
sumed that the author was aware of the modern "principle of vir-
tual velocities," which he probably was not (Lindsay, R. B., ed.,
Energy: Historical Development of the Concept, Dowden, Hutchinson &
Ross, Inc., 1975, p. 32).

48 *Geiger counters.* These are designed to register the passage of particles such as alpha-particles that are emitted during radioactive

decay. A reader of the Australian magazine *Radio and Hobbies* (a monthly periodical on which I and many others cut our electronic teeth) did not know this, and wrote to the editor to ask what Geiger counters counted. The editor, straight-faced, replied: "Why, *Geigers,* of course," and went on to explain that Geigers were what modern scientists counted in order to get to sleep. He even provided a picture of one:

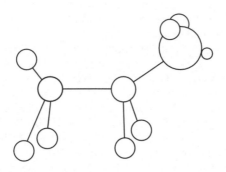

Fig. N.1: A "Geiger."

48 *Geiger and Marsden.* Their observations were published in June 1909, and the whole world had a chance to work out what they meant. Rutherford got there first, although it took him eighteen months.

Rutherford tried *two* pictures of the atom — one where all the positive charge was in the middle, and one where all the negative charge was in the middle. In the first case, the positively charged alpha-particles would be repelled if they came too close. The second case was not ruled out, though, because, although alpha-particles in this instance would be attracted to the nucleus, they could well swing around it and return in the direction from which they came, like a comet around the sun. When Rutherford did the mathematics for the two cases, the answers came out the same. It needed other evidence to show that the first answer was the correct one.

Rutherford's picture of the atom was not just based on one imaginative conceptual leap, important though that was. There have been plenty of equally brilliant conceptual leaps that have landed on the wrong answer. Rutherford, though, used his picture to predict just what fraction of alpha-particles would be scattered through various angles, and found that his quantitative predictions agreed with Marsden's measurements. This was the real clincher.

Rutherford announced his idea, not in the prestigious *Proceedings of the Royal Society* as Geiger and Marsden had done, but to a meeting of the Manchester Literary and Philosophical Society. This says something for the status of such societies and for public interest in real science at that time.

48 *Rutherford and the atom.* Rutherford was fond of recounting the story of Marsden's experiments. The version here is assembled from several sources, the principal one being David Wilson's *Rutherford: Simple Genius*, MIT Press, Cambridge, Mass., 1984, p. 291.

49 *The less likely an experiment is to work, the more significant the result is likely to be.* Unfortunately, the powers now in charge of providing support for science often view things in the opposite light, and pay out largely on the basis of the *a priori* chances of success. This has led to some interesting games, where scientists do experiments first and then apply for the money, using that money to secretly fund the next round.

49 *History of the hammer.* See Goodman, W. L., *History of Woodworking Tools*, Bell, 1964.

My grandfather, a carpenter, claimed to have used the same hammer through his entire working life, and only had to replace the head twice and the handle three times.

49 *Fitting the hammer handle through a hole in the head.* Australian Aborigines used rocks with holes in the middle as spear straighteners, but apparently did not take the next step of fitting axe heads, etc., in a similar way.

52 *Work = force × distance* My search through old scientific articles, books, and encyclopedias has revealed that no one scientist has ever proved that work = force × distance. All that happened was that the conserved quantity (force × distance), identified as important by Galileo, gradually became identified with the word "work," so that by 1855, 150 years after Galileo's death, scientists had hijacked the word from common language for good. Its new definition was accepted without question by scientists such as Joule when he established the equivalence of heat and work.

53 *Conservative forces and friction.* Scientists call the force used to move a load in the absence of friction a conservative force, because the moved load can in principle subsequently be used to then move another one, so that the effort isn't lost. Frictional forces, though, are non-conservative, since the work that has gone into generating heat is usually lost.

54 *The force required to remove a nail.* This is given in *Marks' Standard Handbook for Mechanical Engineers*, 8th edition, McGraw-Hill, 1978,

pp. 12–29. The actual force depends on the 2.5 power of the density of the wood, and is also proportional to the diameter of the nail and the length of embedment.

56 *The length of engagement formula.* This, together with other formulae too numerous and exciting to mention, are given in Ryffel, Henry H. (ed.), *Machinery's Handbook,* 23rd edition, Industrial Press Inc., New York, 1988, p. 1278.

57 *Tightening bolts.* Stuart Burgess has made some interesting further points, which I quote here with his permission:

1. A neat way to tighten a bolt is to heat it up first, and then do it up to hand tightness. When it cools down it shortens, and becomes pretensioned.
2. Many screws are made of soft metals, and are easily damaged by hardened screwdrivers, and still more by hardened misusers of screwdrivers.
3. Spring washers are good at indicating the correct preload in screws and bolts.
4. There is apparently a special screwdriver available with an offset in the middle section of the handle, so that the tightening hand sweeps through a larger circle than just twisting. This does provide a true "lever" mechanical advantage.

63 *The historic definition of a "barrow."* See the complete edition of the Oxford English Dictionary.

63 *The wheelbarrow in China.* This is discussed in Needham, Joseph, *Science and Civilisation in China* (abridged by Colin Ronan), Cambridge University Press, Cambridge, 1978, p. 75.

67 *Cutting tools . . . are the oldest.* Stone flakes discovered in the Nihewan Basin of China, for example, have now been dated as 1.36 million years old (*Nature,* vol. 413, p. 413).

68 *Formulas for Stress and Strain,* Roark, R. J., 3rd edition, McGraw-Hill, 1954.

69 *A screwdriver thus used acts as a rigid extension to the operator's arm.* Stuart Burgess claims that "a screwdriver is really a wrench used in-line." I leave it to the reader to decide between Stuart and Jeff.

69 *Engineering reference information.* Most of this is taken from Baumeister, T., Avallone, E. A., and Baumeister, T., III, *Marks' Standard Handbook for Mechanical Engineers,* 8th edition, McGraw-Hill, 1978.

chapter 4: how to add up your supermarket bill

Page

79 *The principle of concentrating on significant figures.* This is a principle
that scientists use all the time. The trick is to recognize which figures are significant and which aren't. Failure to do this can lead to
scientific ruin, as happened to the congenial Viennese physicist
Felix Ehrenhaft, a frequent host to Einstein and others.

Ehrenhaft spent much of his life trying to measure the charge
on the electron, using a technique where a cloud of tiny water
droplets was sprayed between two horizontal charged metal
plates. Most of the droplets would fall slowly under the influence
of gravity. Occasionally, however, a droplet would pick up a stray
electron liberated by a passing cosmic ray. The charged droplet
would then start to move towards the positively charged upper
plate. By measuring the rate of movement, the experimenter
could calculate the electrical charge. The problem was that some
of the droplets, unaware of the experimenter's intentions, incon-
veniently chose to pick up more than one electron. Ehrenhaft
knew that this was a possibility, and took many measurements,
calculating the charge on each drop and looking for a common
factor which would be the charge corresponding to just one electron. He drew a frequency distribution of his results, showing
how many times each particular value of the charge occurred. His
results for 500 separate experiments are redrawn here as faithfully as I can manage from the full set of data reproduced in G
Holton's excellent book, *The Scientific Imagination: Case Studies*,
Cambridge University Press, Cambridge, 1978, p. 74.

The charge on an individual electron is now known to be 4.80
× 10^{-10} electrostatic units, and it seems obvious in retrospect that
Ehrenhaft's first (and largest) peak corresponded to drops carrying just one electron, while the subsequent peaks (somewhat
displaced because of an unknown experimental artifact) corresponded to droplets carrying two, three, four, etc., electrons. The
scatter in the results (something that all scientists have experienced) could have been due to dust, pairs of droplets sticking together, convection currents in the air, or any of a number of other
reasons.

Ehrenhaft didn't see it that way. He believed that *all* of his results should be given equal weight, since he could see no reason
to believe one more than another, and that they were all accurate to three, four, and even five significant figures, so that each
tiny horizontal step represented, not experimental error, but the

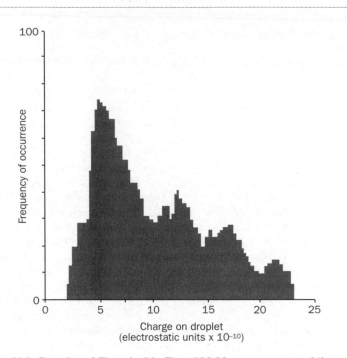

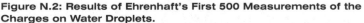

Charge on droplet
(electrostatic units x 10⁻¹⁰)

Figure N.2: Results of Ehrenhaft's First 500 Measurements of the Charges on Water Droplets.

addition of another electron. He thus ended up calculating the charge on the electron to be about a hundred times smaller than it actually is. He eventually performed thousands of experiments. The more experiments he did, the smaller the elementary charge on the electron appeared to become.

The American Robert Millikan was meanwhile performing very similar experiments to Ehrenhaft, but robustly choosing to focus only on the most significant figures and to ignore later figures as being due to experimental variation and therefore not significant. Using this approach, he obtained the correct value for the charge on the electron, for which he was awarded the 1923 Nobel Prize for physics. Poor Ehrenhaft, meanwhile, carried on for another twenty years claiming that there must be "subelectrons" with much smaller charges — but nobody was listening.

There are some very deep statistical issues here, concerning the extent to which prior expectations should influence statistical analysis, and if so just how they should be allowed for. Many of these issues remain unresolved to this day.

80 *Averaging of upper and lower bounds.* This is easily shown to be equivalent to *adding half the number of items to the total in the "pounds" column.* If the total of the "pounds" column is *P,* and the total number of items is *N,* then the lower bound is £*P,* and the upper bound is £(*P* + *N*). The average of the two is £(*P* + (*P* + *N*)) / 2, which comes to £(*P* + *N*/2), which is the same as adding half the number of items (*N*/2) to *P.*

83 *Cancellation of errors.* A surprising number of scientists have come to a correct conclusion after making two mistakes that have canceled each other out. One famous case was when the American scientists Gorter and Grendel were trying to measure the composition of the thin membrane that surrounds all living cells. They eventually concluded (correctly) that this membrane was just two molecules thick. This was a major result, and set modern cell biology on its path. It was only thirty years later that someone pointed out that the Gorter and Grendel paper contained two errors, each of a factor of two, which luckily canceled each other out. See E. Gorter and F. Grendel in *Journal of Experimental Medicine,* vol. 41, 1925, p. 439.

85 *Robert Millikan.* His notebooks are now preserved in ninety-nine file boxes in the California Institute of Technology Archives.

85 *Patterns in supermarket prices.* When I showed these frequency distributions to my colleague Jeff Odell, his immediate suggestion was that I should attempt a Fourier transform. This is a mathematical technique, now made easier with the advent of powerful computers, for revealing underlying patterns, or periodicities. Pictures from space probes are invariably treated in this way before being released to the public.

I was very tempted to try it with supermarket bills, especially since, as Jeff pointed out, there is an experimentally neat way to do Fourier transforms, which is to take a 35 mm slide of the object (here, the graph of the supermarket price distribution) and shine a laser through it. The light will be deflected through different angles via a phenomenon called *interference,* with each angle corresponding to a particular periodicity in the price distribution. In the end, though, it turned out to be too messy.

You can do your own Fourier transform simply by looking at the reflection of a light from a CD. Pick one color to watch, and look for the reappearance of that color as the CD is tilted. Each reappearance corresponds to a different periodicity. Those at the shallowest angle represent the spacing between adjacent tracks; those farther in correspond to the gaps between every second track, every third track, etc.

85 *The underlying statistics.* The statistics of the methods that I have suggested really deserve an essay in their own right, focusing on what we mean when we speak of an "average."

The average that most people are familiar with is technically called a "mean." If the figures in a "cents" column are randomly distributed, it makes sense to say that the "mean" value is approximately 50, for example (49.5, to be exact). It also makes sense to talk about a "mean" if the figures are not randomly distributed, so long as the distribution is smooth.

"Means" correspond to expectations, but are not always appropriate for a situation. The "mean" number of humps on a camel, for example, is 1.5, but no one has ever seen a camel with one and a half humps — they always have either one hump or two humps.

I have avoided talking about "means" when it comes to overall supermarket prices, since the distribution of these prices is very spiky. There are many mathematical tricks for handling such non-smooth distributions, most of which are way out of my league, and probably the reader's as well. One that is within everyone's grasp, though, is to focus on the *median,* i.e., the value that occurs with the highest frequency. This is what I have done in the simple methods that I have suggested for comparing prices between different supermarkets.

89 *The use of calculators.* A friend to whom I was explaining some of the tricks in this chapter asked: "Why not just use a calculator?" My answer was that a calculator is far from a foolproof aid to addition — it is more like a ticking bomb. The bomb becomes primed the moment that a number or a symbol is entered incorrectly. The subsequent explosion may be a personal one, after the operator has used a calculator to add up the same column of figures three times, and produced three different answers. It may even involve other people if the wrong figure is used to challenge someone else's calculation. No professional scientist would dream of trusting the output of a calculator (or a computer, which for most purposes is just a programmable calculator) unless the result agreed reasonably with that arrived at by approximate calculation. Even then, he or she would probably repeat the calculation, just to make sure. There is just too much chance of hitting a wrong key, resulting in GIGO — garbage in, garbage out.

Some spectacular examples of GIGO have been recorded. On 21 July 1962, a misplaced comma in a computer program was sufficient to cause the spacecraft bearing America's first probe to Venus to explode shortly after liftoff. In 1988, the Soviets' first

Mars mission, *Phobos 1*, was lost when Russian controllers sent a long sequence of commands with a single error: a plus symbol where a minus symbol belonged. A single erroneous keystroke is not likely to cost the average person quite as much, but if trained space scientists can make such mistakes, then it behooves the rest of us to be wary.

The easiest way to be wary is to use quick mental shortcuts to check that the figure calculated by you or someone else is at least approximately correct. It is surprising how often such calculations help to avoid paying out on your own or someone else's mistake.

Further examples of GIGO in the space program can be found in an interesting article by Bruce Neufeld called "Software Reliability in Interplanetary Probes," published on the Internet at http://web.tampabay.rr.com/dneufeld/sftrel.html

Belief in the infallibility of calculators, and more so of computers, has reached such a level in some quarters that the meaning of GIGO has been upgraded to "garbage in, gospel out."

chapter 5: how to throw a boomerang

Page

93 *The oldest wooden boomerang.* The oldest wooden boomerang discovered in Australia was found by radiocarbon dating to be 10,000 years old. The age record for boomerangs, though, goes to one made from a mammoth tusk in what is now Poland, and which was dated at 23,000 years.

Boomerangs appear to have been invented independently in many places, including Egypt, where many boomerangs were found in an annex to Tutankhamen's tomb, including some with gold tips.

93 *Modern boomerang materials.* These even include the ultra-strong material Kevlar, used in bulletproof vests and to tether spacecraft, and "borrowed" on this occasion from the brake linings of a Russian MIG fighter by the Bulgarian boomerang enthusiast Georgi Dimanchev.

93 *Aboriginal boomerang "sport."* The account is from Dawson, J., *The Australian Aborigines,* Facsimile edition, AIATSIS, 1881.

94 *The boomerang distance record.* The record of 149.12 meters (long since surpassed) was set by Michel Dufayard at Shrewsbury in 1992.

94 *Invisible boomerangs.* Sean Slade tells the story of the time that he made a boomerang from transparent polycarbonate plastic. He

put some effort into polishing the surfaces and sharpening the edges, and it was only after he launched it that he realized that he had no idea of where it was going to land when it returned. Passersby were treated to the sight of a man running frantically from a totally invisible pursuer.

94 *Boomerangs as hunting weapons.* These are occasionally used for catching ducks, where the boomerang hovers above a flock and appears to the birds as a hawk, driving them towards the ground where they are clubbed.

95 *The best time for maximum time aloft.* This was achieved by John Gorski of the United States with a boomerang that caught a thermal over the Potomac River, returning to within seventy meters of the thrower after an incredible seventeen minutes and six seconds. Unfortunately, the throw was a practice shot, and so the time doesn't count as a world record.

96 *What makes a boomerang come back.* The rationale for this applies with equal validity to boomerangs with more than two arms.

96 *Aerodynamic lift.* A complex subject: the classical explanation is that the air flows faster over one side of a wing (the upper side in the case of an aircraft; the "inner" side in the case of a boomerang) than it does over the other, creating a lower pressure on the side where it flows faster (Bernoulli's principle). The reader can observe this effect by holding two sheets of paper close to each other and blowing between them. The sheets will be drawn towards each other.

This classical explanation has recently been challenged controversially by David Anderson and Scott Eberhardt in a book called *Understanding Flight.* Anderson, a Fermilab physicist, claims that lift is very simple — it's just a reaction force, as described by Newton's Third Law of Motion (for every action there is an equal and opposite reaction). The wing pushes the air down, so in turn air pushes the wing up. I haven't been through Anderson's argument in detail, but it's likely that he's right. This doesn't make Bernoulli wrong. The two explanations produce the same equations — it's only the imagery that is different.

97 *Art on boomerangs.* The decorative imaging has been described by Philip Jones, curator of the collection of over 3,000 Aboriginal boomerangs at the South Australian Museum, in the lavishly illustrated *Boomerang: Behind an Australian Icon,* Wakefield Press, South Australia, 1996.

98 *Aboriginal ignorance of boomerangs.* When I told the story of Duncan MacLennan teaching Aborigines to throw boomerangs, Sean Slade promptly topped it with the story of Andy Furniss, a well-

known British boomerang enthusiast, who was hitchhiking across Australia and had been dropped off in the middle of nowhere. While waiting for his next lift, he idly took out a boomerang and began to throw it. A local Aborigine came up and asked "what the feller thing?" Andy told him, and proceeded to teach him how to throw it. Just as the man was getting the hang of it, a bus full of American tourists drew up and proceeded to photograph what they perceived to be an Aborigine teaching a white man how to throw a boomerang.

99 *Hospital MRI scanners.* These used to be called by the technically correct title of Nuclear Magnetic Resonance scanners. "Nuclear" refers to the nuclei of the hydrogen atoms in the tissue water, but the name was changed when patients (and the media) mistakenly associated "nuclear" with the idea of nuclear weapons, which get their energy from disrupting the nuclei of much larger atoms, whereas MRI scanning gently tilts the single spinning proton at the core of a hydrogen atom.

101 *The Mudgeeraba Creek Emu Racing and Boomerang Throwing Association.* One of its rules reads: "The decisions of the judges are final unless shouted down by a really overwhelming majority of the crowd present. Abusive and obscene language may not be used by contestants when addressing members of the judging panel or, conversely, by members of the judging panel when addressing contestants (unless struck by a boomerang)."

101 *Simplification of physical rules by changing viewpoint.* The most stunning example of this was when German female physicist Emmy Noether pointed out that the law of the conservation of linear momentum arises directly from the fact that the laws of physics do not change when the point of view is shifted linearly in space, and that the law of conservation of angular momentum amounts to no more than the fact that the laws of physics do not change when the point of view is rotated in space. It says something for the masculine orientation that science had in the early twentieth century (and unfortunately still has) that Emmy Noether did not receive a Nobel Prize for this incredible insight.

102 *The mathematics of boomerang flight.* This was spelled out in a semi-popular article by the Dutch physicist Felix Hess (*Scientific American*, 219, 1968, pp. 124–136) and elucidated in excruciating detail by the same author in a 500-page Ph.D thesis published some seven years later. Bob Reid made the physics accessible in "The Physics of Boomerangs" (*Mathematical Spectrum*, vol. 17, 1984/5, p. 48, published by the University of Sheffield), and amplified his account of how boomerangs lie down in a recent edition of the

British Boomerang Society Journal (Summer 1998, p. 19). A more detailed, step-by-step account was also presented by E. C. Zeeman in the splendid Royal Institution Mathematics Masterclass "Gyroscopes and Boomerangs," which was intended "to provide enrichment material for gifted thirteen-year-olds."

104 . . . *eventually resulting in a crazy tumbling.* The mathematical reason for this is that the moment of inertia about the spin axis must be at least 30 percent greater than the next largest moment of inertia about an axis perpendicular to the spin axis, otherwise the spin can transfer from one axis to the other with disastrous results.

104 *The physics of levitating frogs.* This is described by Michael Berry and André Geim in *European Journal of Physics,* vol. 18, 1997, p. 307.

chapter 6: catch as catch can

Page

108 *The description of an English village cricket match.* This is from Macdonell, A. G., *England, Their England* (MacMillan, 1933; The Reprint Society, London, 1941), pp. 122–124. Despite the interval of time, nothing has changed, thank goodness.

I have added a couple of explanatory words (e.g., "the poet" after Mr. Harcourt) to clarify the background without doing violence to the flavor: that of a village life which still exists in spite of everything.

110 *"Running to Catch the Ball"* is a paper by Peter McLeod and Zoltan Dienes, published in *Nature,* vol. 362, 4 March 1993, p. 23.

110 *The brain's unconscious problem-solving abilities.* The psychologist and writer Oliver Sacks tells the story of identical twins who were *savants,* able to "just see" numbers and how they go together. Their ability to visualize numbers was such that, when Sacks accidentally dropped a box of matches, one of them glanced at the scattered pile and said immediately: "A hundred and eleven." His brother promptly commented, "Thirty-seven, thirty-seven, thirty-seven." Sacks counted the matches; there were indeed a hundred and eleven (3 × 37).

The twins' unconscious calculating abilities were extraordinary. They were, for example, able to work out in a few seconds whether or not a given twelve-digit number was prime — something for which scientists still do not have an algorithm (i.e., a set of rules to guide the calculation).

There is thus nothing intrinsically impossible about McLeod and Dienes' postulation of unconscious mental abilities that help us to catch a ball. It is just that the scientist's guiding principle (known in philosophy as Occam's razor) in choosing between two otherwise equal explanations is to accept the simplest. This doesn't mean that the simplest explanation is always the right one; it is just that Occam's razor has turned out from experience to be the procedure that gives us the best chance of selecting the right explanation. McLeod and Dienes' explanation uses the blunt edge of Occam's razor.

110 *Newspaper report of the "ball-catching" equation.* The journalist who managed to get the "ball-catching" equation published on the front page of his newspaper made a brave attempt to explain what the equation meant in lay terms. The result was rather reminiscent of President Kennedy's famous attempt to declare in German: "Ich bin ein Berliner." Not really understanding, he used a word that seemed like the right one, but which came out in translation as "I am a doughnut." Every German in the audience caught President Kennedy's mistake. Only a few of my journalist friend's readers picked up on his mistake, because only a few understood the language of mathematics. That's a pity, because mathematics gives a picture of how things happen, a picture well worth the thousand or so words that a verbal description would take. The description in this case was of how our angle of gaze changes when we run to catch a ball. The equation was described by my journalist as:

$$\frac{d^2(\tan\theta)}{dt^2} = 0,$$ where θ is the angle of gaze, t is the time, and d is the distance

If he had known how to read the language in which the equation was written, he would have realized that distance doesn't come into it at all; d isn't a symbol in a conventional sense, and the expression (d^2 / dt^2) only has meaning if taken as a whole. It means "acceleration" (Newton would have written it by putting two dots above the θ). All that the equation is saying is that, when we run to catch a ball, we judge the catch by running in such a way that our rate of head tilting does not accelerate or decelerate (i.e., it equals zero).

111 *Silvanus P. Thompson, "Calculus Made Easy."* This book has recently been updated and reprinted — see Thompson, S. P., and Gardner, M., *Calculus Made Easy*, St. Martin's Press, New York, 1999.

112 *From the point of view of the observers at the pub door.* The same principle applies to the many satellites that are now in *geostationary* orbit, and which appear to hover above one spot on the Earth's surface. Viewed from outside, they are whizzing around the Earth once every twenty-four hours. The Earth, however, is rotating at the same speed, so that the satellite appears stationary when viewed from a point directly below on the Earth's surface.

113 *Some 30 percent of people still share Aristotle's . . . notion.* Miller, J. D., *Daedalus* 112(2), 1983, pp. 29–48; Durrant, J. R., Evans, G. A., and Thomas, G. P., *Nature* 340, 11, 1989.

Aristotle's was a reasonable, commonsense approach to understanding motion, but flying arrows appeared to constitute an exception, since there was nothing pushing on them. Aristotle was so convinced of the correctness of his approach, though, that he eventually concluded that the bowstring must push the air, even from afar, with the air then pushing the arrow.

Modern-day Aristotelians follow Aristotle in trying, perfectly reasonably, to apply common sense to situations where Nature has decreed that common sense shall not apply. The history of science is littered with such situations, and could, in fact, be said to consist largely of man's attempts to understand and resolve such conflicts between common sense and reality.

113 *Galileo's discovery of the law of acceleration.* This is discussed in Tricker, R. & B., *The Science of Movement,* Mills & Boon, 1968, a book full of absorbing ideas that I found very helpful in writing this chapter. It throws an interesting sidelight on how even the greatest scientists can make mistakes, and also how lucky they can be:

. . . the hypothesis that the velocity of a body increased uniformly with the distance through which it had fallen was the one that was generally accepted. Galileo himself adopted it at first. At the same time he also made a mistake in his early calculations which neutralized the error in this assumption, and he thus arrived at the correct result that the distance that a body would fall in a given time would be proportional to the square of the time . . .

(He later saw the error in his earlier calculation, and also realized that the "uniform velocity" hypothesis was wrong).

113 *The formula for a parabola.* This takes the form (vertical distance) = (constant) × (horizontal distance)^2. Since a projectile travels horizontally at a constant speed, the horizontal distance traveled is proportional to the time of flight, so that the equation can be writ-

ten (vertical distance) = (different constant) $\times$ (time)^2, i.e., the distance fallen is proportional to the square of the time.

A parabola is one of the "conic sections" originally described by the Greek mathematician Apollonius of Perge. Their shapes can be obtained by slicing a cone at different angles. One of these sections is the circle (obtained by slicing a cone parallel to the base). Another is the ellipse (obtained by slicing the cone at an angle to the base). The parabola is a third, obtained by slicing a cone parallel to the side.

114 *It is not a cue that is always reliable.* As the mathematics eventually showed, it is the tangent of the angle that needs to change at a constant rate, rather than the angle itself. Our childhood cue of using the angle, rather than its tangent, works well because an angle is proportional to its tangent for small angles, and is actually nearly equal to its tangent if the angle is expressed in "natural" units, rather than the more familiar degrees, which arbitrarily divide a circle into 360 equal parts and which derive from a Babylonian counting system based on the number 60.

The "natural" unit is the radian. An angle of one radian corresponds to a wedge of a circle where the length of the circumferential arc is the same as the length of the radius (Figure N.3).

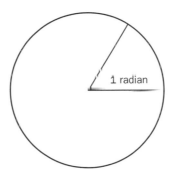

Figure N.3: An Angle of One Radian.

The table below shows how close the tangent of an angle is to the angle itself, when the angle is expressed in radians:

Table N.1: Closeness of Angles and Radians.

Angle (degrees)	Angle (radians)	Tangent of angle
10	0	0
10	0.18	0.18
20	0.35	0.36
30	0.52	0.58
40	0.70	0.84
50	0.87	1.19
60	1.05	1.73
70	1.22	2.75

The error in the approximation is around ten percent or less for angles up to 30°, but rapidly worsens for larger angles, which may partly explain why we find it so difficult to judge high catches.

115 *The "ball-catching" equation for the case of a catcher standing in the right place.* The equation (tànα = $g/2v$) leads directly to that published by McLeod and Dienes (d^2 (tanα) / dt^2 = 0), since the expression ($g/(2 \times v)$) has a constant value for any particular catch, so that the rate at which d (tanα)/dt itself changes with time (i.e., d^2 (tanα)/dt^2) is zero.

This was also the form in which the equation was originally published by S. Chapman (*American Journal of Physics* 36, 1968, pp. 868–870) and P. Brancazio (*American Journal of Physics* 53, 1985, pp. 848–855). Both of these references are quoted in the subsequent *Nature* paper. It was probably the difference in form that led McLeod and Dienes to miss the obvious physical interpretation that I have discussed in this chapter.

118 *Action and reaction.* This is Newton's Third Law of Motion ("For every action there is an equal and opposite reaction"). In other words, if we push on something, it pushes back just as hard. From the "something's" point of view (e.g., the ground) it is pushing on us, and we are pushing back just as hard. It just depends on your point of view.

chapter 7: bath foam, beer foam, and the meaning of life

In this chapter I have tried to give a picture of how science actually works, which is not usually by preplanned strategies but more often by a combination of awareness of what questions are important, a readiness to take up new ideas that might be relevant to those questions, and sheer, dogged persistence in following up the consequences of these ideas.

In order to give the detail, I have focused on work where I have had direct knowledge of what was going on, and even then I have been able to mention only a few of the scientists concerned. I have reported most of the incidents from the point of view of how I saw things at the time, and I was not always privy to everything that was going on. Others may well have seen things differently, or with a different emphasis. In other words, to quote from Lawrence Bragg's introduction to *The Double Helix*, "this is not a history, but an autobiographical contribution to the history that [may] someday be written." (Bragg was the head of the Cavendish laboratory during the period when James Watson and Francis Crick devised their double helical structure for DNA [1951–1953]. Watson, J. D., *The Double Helix*, Atheneum Press, New York, 1968).

Page

121 *How molecules self-assemble.* A good starting point for non-specialist scientists who wish to follow up is the articles in *Current Opinion in Colloid and Interface Science* (February 1999).

121 *American Scientist Sidney Perkowitz.* On the BBC radio program *Lord Kelvin's Bedspring,* Sidney Perkowitz and I discussed the science of foams at some length. Those pundits who believe that the public is interested only in the showier aspects of science might like to note that this program, dealing with such an apparently mundane subject, became BBC Radio 4's "Pick of the Week."

122 *Sir Isaac Newton.* Newton published his observations of the colors of soap bubbles in *Opticks* (G. Bell & Sons, London, 1730), a wonderfully readable account of his ingenious (though incorrect) corpuscular theory of light.

123 *. . . about ten times thinner than can be observed with the naked eye.* Even with the aid of a powerful microscope, the human eye can only resolve objects that are more than a micrometer or so apart. It can detect the presence of much smaller objects, though, through their ability to scatter light, which is why car headlights stand out so vividly in a fog, even though the observer can't make out the shapes of the individual water droplets.

123 *Irving Langmuir.* Langmuir's very interesting biography is given in *The Selected Papers of Irving Langmuir*, vol. 12, Pergamon Press, 1962.

123 *Stories surrounding Irving Langmuir.* These are many and varied. He looked for science in everything about him, and was never without a notebook and measuring tools. On one famous occasion he used the entire toilet paper supply of a remote guest house where

he was staying in pursuit of an experiment. He was also a keen amateur pilot, and fond of playing a game which involved close encounters with clouds. On one occasion he passed his wheels through the top of a cloud, and was astonished to find that the wheel-tracks remained. In following up this observation he eventually developed new methods of weather modification, mostly concerned with "cloud seeding" by dropping dry ice into the cloud to initiate precipitation.

I once met a scientist who had worked down the corridor from Langmuir, and who told me that Langmuir's laboratory was a constant source of surprises. On one occasion this scientist walked into it, only to find himself enveloped in a snowstorm, artificially created by Langmuir and his colleague Vincent Schaefer as part of their experiments on weather modification.

126 *Liposomes formed spontaneously from lecithins manufactured naturally under the conditions that then prevailed on Earth.* The destructive oxidizing nature of the atmosphere of the early Earth has led to some speculation that complex organic molecules could not have been formed there and were actually "seeded" via carbonaceous chondrites, such as the Murchison meteorite discovered in central Australia. My colleague Professor Richard Pashley has found surface-active materials in the core of this meteorite, which is dated 200 million years older than the Earth itself (Deamer, D. W., and Pashley, R. M., "Surface properties of amphiphilic components of the Murchison carbonaceous chondrite," *Origins of Life and Evolution of the Biosphere,* 19, 21–38 (1989).

127 *Liposomes as precursors for living cell membranes.* This notion was first proposed by Deamer and Oró (*Biosystems,* vol. 12, 1980, p. 167).

127 *Alex's authoritative textbook on colloids.* The book was modestly entitled *Colloid Science* (A. E. Alexander and P. Johnson, Oxford University Press, 1947).

128 *The adhesive properties of sickle cells.* These were studied by Evan Evans at the University of British Columbia in Vancouver. Evan, a former engineer, turned his attention to the mechanical properties of living cells, where he rapidly became the world authority, and much feared at conferences for his devastatingly straightforward questions and refusal to accept evasive answers.

In his studies on sickle cells, Evan found that their inability to recover from deformation was an important factor, perhaps more important than their adhesiveness.

128 *The DLVO quartet.* Theo Overbeek is the only one still alive. He was still scientifically active at the time of writing — a remarkable record.

129 *The repulsive forces between charged head-groups.* These actually arise from the overlap of the clouds of charged ions that gather about each head-group. For this reason, the forces are technically called "double layer" forces. Foam stability is not a matter of DLVO theory alone. One Dutch scientist showed me this when he took me on a tour of Amsterdam bars, pointing out that some of the beers kept their heads very well, while others collapsed much more quickly. The difference, he had discovered, was that the more rapidly collapsing froths contained "natural" carbon dioxide from the brewing process, while the longer-lasting ones had been produced with nitrogen from a gas cylinder. Carbon dioxide, unlike nitrogen, dissolves in water, and can hence escape into the atmosphere through the water films surrounding the bubbles, so that the bubbles in a "natural" froth shrink rapidly with time.

I kept meeting this scientist at conferences over the next twenty years. He was often accompanied by his wife, a lady of rather forbidding appearance, who was never more so than at a conference in Bulgaria, when the local organizers produced a troupe of nubile dancing girls for our entertainment at the conference dinner. Her husband was fascinated, but his wife appeared to be rather less so.

131 *Jacob Israelachvili.* The work of Tabor and his school was summarized by Israelachvili in his excellent book *Intermolecular and Surface Forces,* Academic Press, 1985. The term "surface forces" simply refers to the close-range attractive and repulsive forces between surfaces whose interplay dominates many of the processes of life, including self-assembly. Jacob points out in his introduction that the Greeks needed only two such forces to account for all natural phenomena. One was Love, and the other was Hate.

132 *Barry Ninham.* His book (with Mahanty) on Van der Waals forces is Mahanty, J., and Ninham, B. W, *Dispersion Forces,* Academic Press, 1978.

133 *Louis Pasteur.* As a scientist, Pasteur is mostly famous for having discovered the principles of sterilization (which led, among other things, to the process of "pasteurization" and of vaccination using attenuated (i.e., live, but weakened) vaccines). Going against the advice of colleagues, he dramatically tried his untested rabies vaccine in 1885 on a nine-year-old boy, Josef Meister, who had been bitten by a rabid dog. Meister lived to become caretaker of the Pasteur Institute, and died fifty-five years later by committing suicide rather than open the tomb of Pasteur to invading Nazi forces.

133 *Pasteur's discovery that molecules have three-dimensional structures.* When Pasteur discovered this he wasn't trying to understand molecular shape, any more than the scientists who initiated the

foam revolution were trying to understand the origins of life. It was an insight that came, as so often in science, from an attempt to answer a totally different question.

The question was posed by the French wine industry, whose members wanted to know whether the newly discovered racemic acid, a by-product of some industrial processes in the Alsace region of France, was the same as tartaric acid (the main component in the "crust" thrown by a good wine).

133 *Twisting a beam of light.* More technically, the tartaric acid solutions that Pasteur studied rotated the plane of polarization, while the apparently identical racemic acid solutions did not. Pasteur traced the difference to the fact that racemic acid is in fact a mixture of two acids, one of which is tartaric acid and the other of which has molecules that are mirror images of those of tartaric acid, and which rotate the plane of polarization in the opposite direction to those of tartaric acid, so that in a mixture there is no net effect.

If Pasteur had been a cuttlefish, he could have seen these effects directly, since cuttlefish eyes are adapted to respond to the rotation of a beam of light that has passed through otherwise invisible prey, such as the glass shrimp. As it was, Pasteur had to use a relatively new invention called a polarimeter. A rough polarimeter can be made by taking two pairs of (genuine) Polaroid sunglass lenses, placing one behind the other, and looking through both together at a bright light. If one is rotated, a point will be found where the light is practically cut out. This is because the first lens selects light waves vibrating in one direction, but the second is set only to pass light waves vibrating at 90° to that direction, and to cut out all light waves vibrating in the original direction.

If a thick piece of clear plastic, or a concentrated sugar solution, is placed between the two lenses, the light will reappear because both of these materials rotate the plane of polarized light.

133 *My one and only paper on the Kerr effect.* This was published in the *Journal of the Chemical Society*, 1963, p. 4450.

134 *The scanning probe microscope picture.* This was kindly provided by my Bristol University colleague Professor Mervyn Miles, a world authority on the imaging of biological molecules by this technique, and a remarkably fine pianist at the annual departmental Christmas show.

135 *The Franklin quotes.* These are taken from his collected papers. The pond where he did his experiments on Clapham Common was called Mount Pond, which was dug by his friend (and banker!) Henton Brown.

135 *Franklin's letter to William Brownrigg.* The letter was transmitted by

his friend to the Royal Society, which promptly published it (with slight modifications) in *Philosophical Transactions of the Royal Society,* vol. LXIV, 1774, p. 447. Would that it was as easy to get a piece of work published these days.

135 *Pouring oil on troubled waters.* This idea goes back to the Venerable Bede, who gave this sage advice in his *Ecclesiastical History,* book 3, chapter 15 ("Remember to throw into the sea the oil which I gave you, when straightway the winds will abate, and a calm and smiling sea will accompany you throughout your voyage"). Unfortunately, it doesn't work with waters that are not so much troubled as psychotic. My Australian contemporary, Bill Mansfield, found this in the 1950s when he repeated Franklin's experiment with the aim of using the surface film to reduce the rate of evaporation of water from Australian dams. Instead of olive oil, he used a waxy surface-active material called stearic acid. Stearic acid, unlike olive oil, will spread indefinitely on a water surface unless confined, but Bill's idea was to put enough stearic acid on the dam surface to form a packed layer of molecules, with the walls of the dam limiting the spread. It was a clever idea that worked well in small-scale experiments, but which failed on a larger scale because high winds caused ripples large enough to break up the stearic acid layer and exposed the underlying water. If the idea had succeeded, it would have been a remarkable, almost pollution-free achievement. Stearic acid is a natural molecule that degrades fairly easily, and in any case a layer of stearic acid containing sufficient material to cover a dam would still contain only a few grams of material.

135 *Lord Rayleigh's bathtub experiments.* These were reported in *Proceedings of the Royal Society,* vol. XLVII, March 1890, p. 364.

136 *Agnes Pockels.* Her experiments were reported in *Nature,* vol. 43, 1891, p. 437.

136 *The first description of a Langmuir trough.* This appears to have been given by Langmuir in *Journal of the American Chemical Society,* vol. 39, p. 1848.

138 *... molecules will get as far away from each other as possible, like relatives at a wedding.* The repulsion between the electrically charged head-groups usually outweighs the Van der Waals attractive forces between the tails at all distances.

138 *The single layer of molecules on the surface of a Langmuir trough.* The molecules can actually be picked up on a glass slide that is lifted up through the surface. A second layer can be picked up on top of the first one by passing the slide back down through the surface, and the process repeated indefinitely so long as there are

molecules left to pick up. This technique, originated by Irving Langmuir and Katherine Blodgett, is at least known by both names and is called the Langmuir-Blodgett technique. It could in theory be used to make "thin film" electronic devices with unique characteristics, but despite many efforts, small imperfections in the films have proved to be an almost insuperable problem.

138 *Cholesterol.* This substance is an ideal candidate for the Langmuir-Blodgett technique, which I used early in my career to make multiple layers of cholesterol that showed brilliant interference colors in reflected light.

138 *The early uses of the Langmuir trough to measure molecular shapes.* These were summarized by Alex in a review (*Annual Reports of the Chemical Society,* vol. 41, 1944, p. 5).

140 *Margaret Thatcher's contribution to science.* This is recorded by H. H. G. Jellinek and M. H. Roberts in *Journal of the Science of Food and Agriculture,* vol. 2, 1951, p. 391. Margaret Thatcher's political career took her to Parliament, where, as leader of the Conservative Party, she served as prime minister for a number of years, overlapping with Ronald Reagan, for whom she had a great (and mutual) admiration.

140 *Denis Haydon.* Haydon described his early work on BLMs with his many coworkers in a typically thorough review in *Methods in Membrane Biology,* vol. 4, 1975, p. 1. He became a father figure, or at least an uncle figure, to the many students who passed through his laboratory, and stories about his natural inclination to take the lead were legion. One that he used to tell against himself concerned his favorite sport of rock-climbing, in which he was indulging alone on the Isle of Skye. Camped at the base of a cliff, he fell into conversation with a fellow camper, and offered to lead him on a climb. The fellow camper, it appears, was no climber, but Denis had a very persuasive way with him and, as I found when working with him, would not usually take no for an answer.

They set off the next morning with Denis leading and his reluctant follower roped on behind. Denis had to do most of the work and was relieved when he made it to a high ledge where he sat, waiting for his follower to appear. He waited for some time, and eventually pulled on the slack rope, to discover that it was supporting a dead weight. Alarmed, he began to pull on the rope, and after some time the other end appeared. It was tied to a huge boulder. When Denis peered over the edge of the ledge, he saw the small figure of his erstwhile follower running off into the distance. I often use this story as a metaphor for what it feels like to be a scientist.

Denis died of bone cancer some years after the publication of our results. In his *Biographical Memoir* for the Royal Society (*Biographical Memoirs of Fellows of the Royal Society,* vol. 36, 1990, pp. 199–216), he described me as "persistent" — a true compliment to a scientist.

141 *Josef Plateau's* original papers are unfortunately hard to obtain. I eventually tracked them down in the *Annual Reports of the Smithsonian Institution* for 1864–1866!

141 *Plateau's results.* These, together with many other fascinating facts about the science of soap bubbles, were popularized in a series of packed lectures "intended for juveniles" by a remarkable Victorian scientist called Charles Vernon Boys, inventor of the high-speed camera. Boys, apart from being a top-class experimental scientist, was also one of the first scientific humorists, whose activities included the creation of giant smoke rings that he dropped over unsuspecting passersby from the window of his laboratory on the first floor of London's Royal Institution. He is most famous among surface scientists for persuading the Society for the Promotion of Christian Knowledge that the book of his lectures, *Soap Bubbles and the Forces that Mold Them,* was a suitable subject for their imprint. I am fortunate enough to possess a first edition of this book, which was published in 1890 (the year that Rayleigh published the results of his "bathtub" experiments), and which is the only book I know that is dedicated to a school science master.

141 *Denis Haydon's work on anesthetics.* See, for example, D. W. R. Gruen and D. A. Haydon in *Pure and Applied Chemistry,* vol. 52, 1980.

142 *The phenomenon that keeps oil and water molecules apart.* This is known as the "hydrophobic effect," and has been the subject of an intense amount of research. Technically, it occurs because the entry of an oil molecule into liquid water forces the water to adopt a more ordered structure, decreasing its entropy, a process which is the opposite of spontaneous. How this happens, though, remains a mystery. The trouble is that the effect on the energy of the system is composed of several different changes, each of which is huge, but which go in different directions, so that the overall effect is tiny. To calculate that tiny effect, though, each of the huge changes must be known to incredible precision, still way beyond the power of the present generation of computer models.

143 *The seminal "molecular packing" paper.* This was published by Jacob Israelachvili, John Mitchell, and Barry Ninham in the *Journal of the Chemical Society,* Faraday Transactions, vol. 72, 1976, p. 1525.

144 *Barry Ninham's work on microemulsions.* See, for example, Zemb, T.

N., Barnes, I. S., Derian, P. J., and Ninham, B. W., *Faraday Transactions of the Royal Society of Chemistry*, vol. 81, 1990, p. 20.

144 *Commercial dishwashing liquids.* Such liquids are formulated to
clean well, but also to foam. The foam is stabilized by molecules
with a different shape than those which contribute to the cleaning, and is only there as an optional extra, designed to impress the
consumer with the efficacy of the product, but stabilized by molecules whose shapes mean that they do little of the actual work.

144 *The interaction between two BLMs.* This was reported in the *Royal Society of Chemistry Faraday Discussion*, no. 81, 1986, p. 249.

145 *The saving of the human race.* For example, it may be possible to pre-
serve our genetic record for future reproduction by encasing DNA
in artificial bilayers produced by self-assembly.

chapter 8: a question of taste

Page

147 *Brillat-Savarin.* One of the best-known names in gastronomy. His
remarkable book, whose title is usually shortened to *Physiologie du
goût,* was published shortly before his death. *"Goût"* (pronounced
"goo") is an untranslatable French word that reflects the full flavor experience, which is more than just a combination of taste
and aroma. Its untranslatability is verified by the fact that my
computer spellchecker, with unintentional humor, keeps correcting the spelling to "gout."

It may say something about cultural attitudes to food that Brillat-
Savarin's book was not translated into English until 1884 (the
American food writer M. F. K. Fisher's 1949 translation is now
generally regarded as the best). If it were not for this book, his
name would live on only in the dish called a Brillat-Savarin,
which is a method of serving lamb in small pieces, accompanied
by duchess potatoes, foie gras, truffles, and green asparagus tips in
butter. I find myself salivating at the very thought of this dish,
which is another example of the fact that good food is as much a
matter of expectation as experience.

Brillat-Savarin's reverence for the aftereffects of eating were reflected by Rossini, who was no mean gourmet himself, in four
musical pieces collectively entitled *The Gourmet Life* and intended
to represent the process of digestion. Their subtitles were, in
sequence, *Interrupted Contentment, Over-Indulgence, Juices* (a piece
which combined the sound of hardworking gastric juices with
some repetition in unexpected places), and *Relief.* I believe that I

possess the world's only recording of this little-known set of pieces, kindly made privately for me by the British composer Malcolm Hill.

149 *The Futurists.* See Marinetti, Filippo Tommaso, *The Futurist Cookbook*, Brill, S., transl., Bedford Arts, 1989. A massively entertaining account of "Food as a Performance Medium" was published by Barbara Kirshenblatt-Gimblett in *Performance Research*, volume 4, page 1 (1999).

149 *Heston Blumenthal.* Blumenthal is chef-proprietor of the Fat Duck restaurant at Bray (near Windsor), in the U.K. I am indebted to him for much information and many enjoyable conversations about cooking styles.

149 *. . . garlic and coffee.* Professional flavorists have a similar concept of "bridging the gap" between two dissimilar aromas to improve the overall impression.

149 *the brain . . . can't decide whether it is experiencing garlic or coffee, and oscillates between the two.* This "explanation" is admittedly speculative, and Gary Beauchamp, director of the Monell Chemical Senses Institute, has cast some doubt on it.

150 *. . . the human brain loves surprises.* Experiments at Emory University Health Sciences, reported on *www.sciencedaily.com.* Perhaps an alternative interpretation is that the brain hates boredom and cuts out signals that have been around too long. This obviously has survival value when we are constantly bombarded by lots of different aromas. I am indebted to my friend and colleague Dr. Alan Parker, of Firmenich Plc, for making this and a number of other interesting points about this chapter

151 *. . . flavor scientists are making progress.* A great deal of exciting work in this area is emanating from the Monell Chemical Senses Institute in Philadelphia, led by Gary Beauchamp, and the research laboratories of Firmenich Plc in Geneva, Switzerland, led by Tony Blake, a fount of information on many things sensual.

151 *. . . putting a few coffee beans under the grill.* Another way to fool people with instant coffee is to add a little ground cardamom, filtering off the residue before serving.

152 *The gravy project.* This was sponsored by Bisto, makers of a range of instant gravies. The difference between such commercial gravies and those prepared by cooking a little flour in meat juice lies mainly in the fact that the starch usually comes from a different source in the commercial gravy, and the commercial gravy is fat-free, an advantage for health, but a disadvantage when it comes to flavors that only dissolve in fats.

152 *Peter Barham.* Barham is the author of *The Science of Cooking*

(Springer-Verlag, 2001), and is also the person who introduced me to the wonderful world of gastronomical science.

153 *The Gravy Equation:*

$$\% \text{ Gravy Uptake} = \frac{(W - (D/S))}{D} \times 100$$

where W is the wet (or uncooked) weight of the food, D is the dry (or cooked) weight, and S is the shrinkage factor. I published the details in "The Theory of Gravy," *Annals of Improbable Research,* vol. 7, 6, November/December 2001, p. 4.

153 *One food that didn't make it to the final gravy report.* Popcorn was the most porous food material that we could think of, and therefore the one likely to take up the highest percentage of gravy. It did — it took in six times its own weight. Wendy opined that the taste of gravy and popcorn is disgusting, and that individual Yorkshire puddings, which take up 90 percent of their own weight, were much to be preferred. She was very disappointed when we threw the puddings out after we had weighed them. Peter had to cook some more.

155 *. . . unanswered questions.* I have not had a chance to return to these questions, although I have passed the information on to a few chefs, and also published the data in the aptly named *Annals of Improbable Research.*

155 *The Devil's Dictionary. The Unabridged Devil's Dictionary,* Bierce, Ambrose, Schultz, D. G., and Joshi, S. T., eds., University of Georgia Press, 2002.

156 *The X-ray video of a chewing head.* The video was shown by Professor Robin Heath of the Royal London School of Medicine and Dentistry. Robin makes a particular study of problems in eating, which are surprisingly common, especially among older people. The main problem, he tells me, is an increasing inability to secrete sufficient saliva, a problem often exacerbated by particular drugs.

Robin did a survey of which food patients found easiest to eat, and was surprised to find that the hard and brittle gingersnap cookie came high on the list. These cookies can support a weight of five kilograms when suspended across a gap, so Robin had every right to be surprised. He eventually discovered that none of his patients had thought to mention that they dunked the cookies in tea to make them softer before eating them.

158 *"shear-thickening" and "shear-thinning."* The bolus becomes more coherent as we chew, which makes it harder to distort. The terms that I have used to describe this have a more precise scientific meaning, which may or may not strictly apply.

159 *this part of the tongue . . . first encounters the lactose (one of the sweetest of sugars) in mother's milk.* This is only true if the infant licks before suckling. Once suckling starts, the milk is expressed to the back of the mouth.

165 *Menthol modulates oral sensations of warmth and cold.* Details of this can be found in Green, Barry, *Physiology and Behaviour,* vol. 35, 1985, p. 427.

165 *Chili pepper hotness.* This is measured in "Scoville units," a measure of the concentration of capsaicinoids. The popular jalapeno chilis weigh in at up to 5000 Scoville units, but for real heat try the waxy, thumb-sized habanero (up to 300,000). Pure capsaicin registers a staggering 15 million.

Like many of the aroma molecules, capsaicinoids are soluble in oil, but insoluble in water or cold beer. For the non–chili pepper addict, or the addict who has had enough, the secret to clearing chili pepper heat from the mouth is to drink plenty of milk. Coconut also works because it is loaded with oil. Overconsumption of chili peppers, incidentally, can lead to hemorrhoids.

167 *The collapse of an aqueous draining film.* My results for this experiment were reported in *Colloids and Surfaces,* vol. 52, p. 163.

167 *The "spraying" of droplets containing aroma molecules.* The original idea that aroma droplets might somehow be "sprayed" from a food bolus was suggested to me by Professor Robin Heath.

chapter 9: the physics of sex

Page

171 *I was once asked to give a talk.* The aim of my talk, and of this chapter, was to make science accessible and to use the journey of the sperm to show that science is not divided into watertight compartments labeled "physics," "biology," "chemistry," etc. It was not my intention to give scientific answers to sexual problems, although a surprising number of people have asked me to do just that. My answer has always been that I am not qualified in this area, and have never worked in it professionally. The best I can offer is information that people may be able to use to help them to formulate the right questions to someone suitably qualified.

171 *There were more teachers than students present.* Middle-aged male teachers may have had a special interest. According to my Bristol University colleague, Professor Shah Ebrahim, middle-aged men who have sex three or more times a week are half as likely to have strokes or heart attacks as those who are sexually inactive (con-

tinuing studies on 2400 Welsh men, reported at the World Stroke Congress in November 2000).

Ebrahim's results are in stark contrast to the belief of John Harvey Kellogg, inventor of Kellogg's Cornflakes, that those who engage in sex, even for procreative purposes, should limit their activities or insanity would result (Kellogg, J. H., *Plain Facts for Old and Young*, Senger, Burlington, Iowa, 1882).

171 *The radioactive watch and telephone book stories.* These were reported in a survey for *Doctor,* a weekly newspaper for the British medical profession (quoted in *The Editor,* 21 July 2001).

Reports of teenage misunderstandings about sex are endless. Two of my favorites come from Miami. One pregnant teenager attending a Miami Beach clinic insisted that she had suffered from contraceptive failure. "What method did you use?" she was asked.

"Jelly," she replied.

"What type of jelly?"

"Grape."

The same clinic gave a seventeen-year-old boy a demonstration of how to use a condom. "I've been doing it wrong for two years," he said. "I thought you had to poke a hole in it so your testicles wouldn't explode."

172 *Sex is even now regarded as a somewhat dubious topic for a scientist to be discussing.* A history of sex research is given by the American academic Vern Bullough in *Science in the Bedroom* (Basic Books, 1994). Professor Bullough claims that a great deal of the stigma still associated with sex research goes back to St. Augustine's doctrine "that the sin of Adam and Eve is transmitted from parents to children through the sexual act, which, by virtue of the lust that accompanies it, is inherently sinful." My own belief is that a great many people in the Western world still subscribe to this doctrine implicitly, even if they do not do so explicitly.

Some of the people at the talk to which I referred may have been a little shocked when I introduced a study by Swedish scientists who persuaded a couple to make love in a hospital MRI scanner of the type used to study brain abnormalities. In this case, the couple's sexual organs were the focus of attention. One can only admire their dedication to duty as they managed to have intercourse while being bombarded by a series of instructions through a microphone.

173 *Blood pressure generated by pumping of the heart.* The blood pressure is kept within "normal" limits by the elastic stretching of the blood vessels with each pumping stroke. If it were not for this, the systolic blood pressure (the higher of the two readings normally quoted) would be incredibly high, because liquids are virtually in-

compressible, as anyone who has ever belly-flopped from a high diving board will know. One of the causes of high blood pressure is in fact a gradual stiffening of the blood-vessel walls so that they are unable to dilate sufficiently.

173 *Hormones . . relax the smooth muscle of the artery walls.* In fact, they stimulate the release of nitric oxide, a small molecule whose chemical symbol (NO) rings rather oddly in this context. It is nitric oxide that produces the relaxation. The nitric oxide is eventually destroyed by adrenaline, or men would be sporting day-long erections.

174 *Vacuum device for passive erection.* Researchers at the Novosibirsk Research Institute in Russia claim that a course of this treatment, combined with the use of an infrared laser to irradiate the top of the penis for five minutes each time, eventually leads to a situation where the vacuum device is no longer necessary because the laser treatment produces "a massaging of the veins which leads to increased metabolism and nutrition of the tissue."

174 *There is no such thing as an aphrodisiac.* There are, however, materials that can produce orgasms in some people. The antidepressant clomipramine makes some women have an orgasm when they yawn or sneeze (*New Scientist,* 27 March 1998, p. 27). A friend of one sufferer asked what she was taking for it. She answered, "Pepper" (*New Scientist,* 20 June 1998, p. 56).

Other antidepressants have the opposite side effect. At least one-third of people taking antidepressants along the lines of Prozac suffer a loss of libido or have difficulty attaining orgasm (*New Scientist,* 29 September 2001 p 17).

174 *Fancied resemblance to a penis or a vagina* The range was fairly wide. In the words of Piet Hein's little "grook" (Hein, P., *Grooks,* MIT Press, Cambridge, Mass., 1966):

Everything's either concave or -vex,
So whatever you dream will be something with sex.

174 *Aphrodisiacs.* Ginseng has recently been shown to *reduce* the number of sperm entering the cervical mucus.

One of the more unusual "aphrodisiacs," described in the *Kama Sutra,* consists of a powdered mixture of the dried plant *Vajnasunhi* (its Sanskrit name — I have been unable to find the English equivalent), red arsenic (As_2S_2), and sulphur. The mixture is set on fire. If the moon, viewed through the exceedingly poisonous smoke, appears golden, then the amatory experience will be successful. There is some fascinating physics involved here, since the moon will appear golden only if the smoke particles are the same size as the wavelength of blue-green light (about 500 nm), so that they will scatter this light but leave the red/yellow end of the spectrum relatively unaffected.

The mixture is also unusual in that it can be turned into an *anaphrodisiac* by adding monkey excrement and throwing it over the maiden. This ensures that she will not be given in marriage to anyone else, which is hardly surprising.

174 *Spanish fly and French troops.* Reported by Meynier, Dr. J., *Archives of Military Medicine and Pharmacology,* vol. 22, 1893, p. 52.

175 *"Cupid's Nightcap."* New Statesman and Nation, 7 November 1953.

176 *Effects of Viagra on cut flowers.* British Medical Journal, vol. 313, 1999, p. 274.

The FDA site for consumer information on medical aspects of Viagra is: http://www.fda.gov/cder/consumerinfo/viagra/default.htm.

Viagra acts by slowing the degradation of nitric oxide. It is not the only orally administered drug that can help sufferers of impotence. Losartan, a drug used to combat high blood pressure, was found in a clinical trial to help nearly ninety percent of sufferers (Capo et al., *American Journal of Medical Sciences,* vol. 321, 2001, p. 336).

176 *The physics of ejaculation.* I have used the example of ejaculation with some success to teach Newton's Laws of Motion to an otherwise reluctant class of first-year university students. The speed with which the ejaculate emerges can be worked out by the application of Newton's Second Law of Motion (force = mass × acceleration), which can be used to calculate the vertical speed of a projectile from the maximum height that it attains. In this case the "projectile" is a teaspoonful of ejaculate. The only published description I could find about the height it can attain appears in Philip Roth's *Portnoy's Complaint,* where Portnoy is reported to have hit the light bulb. Information from other sources, however, revealed that the average male is unlikely to be able to launch his teaspoonful of ejaculate more than a foot or so straight up into the air. In this case, Newton's Second, Law says that its initial velocity must be around two meters/second, i.e., around seven kilometers per hour, which is a fast walking speed.

This doesn't create too many problems on Earth, but it does add an interesting twist to the problem of making love in space. The problem is not just an academic one — NASA has now begun to issue pregnancy testing kits to female personnel on the International Space Station, thereby admitting that a group of men and women cooped up together for five months in space might get up to more than a bit of meter-reading and dial-twisting.

According to NASA technician Harry Stine (in Stine, G. H., *Living in Space*, M. Evans & Co., 1997), NASA has already conducted ground-based experiments on the feasibility of making love in space. The experiments were conducted in a buoyancy tank, although the names of the volunteers are unfortunately not on public record. The conclusion was that making love in weightless conditions is barely feasible, but that it is made much easier if a third person is present to hold one of the bodies in place.

Whales found this out eons ago, and many species of whale use the practice to this day. In physical terms the third person (or whale) is there to provide a defense against Newton's First Law of Motion, which says that if you push on something, it will accelerate away unless there is a balancing force that stops it from doing so.

In space, there is another of Newton's Laws to be considered — his Third Law, which says that action and reaction are equal. This principle explains why a shell fired from a cannon produces a recoil that drives the cannon backwards. The principle applies equally well to gases and predicts, for example, that burning fuel ejected from a rocket will produce a recoil that drives the rocket forward. It also applies to liquids, including those produced during ejaculation.

The exact amount of recoil can be calculated from the principle of the conservation of momentum, which says that the forward momentum (i.e., mass × velocity) of the ejaculate must be balanced by the backward momentum of the body ejecting it.

According to my calculations, if an eighty-kilogram man ejects three grams of ejaculate traveling at seven kilometers per hour, he will recoil at an initial speed of $(0.003 \times 7/80) = 0.00026$ kilometers per hour. In a gravitational field this doesn't matter too much. If the sperm is directed downwards, for example, the man will recoil upward by a maximum distance of five micrometers before being brought back to earth under the influence of gravity.

In space, though, the man will keep moving, covering one meter every three hours, until he hits one of the walls of the spaceship. If the spaceship is eight meters long, he could take as long as twenty-four hours to reach the far wall — just nice time for the libido to build up for a return bout.

177 *Chances of conception after discontinuation of intrauterine and oral contraception.* These were reported by C. Tietze in the *International Journal of Fertilization*, vol. 13, October–December 1968, p. 385.

177 *. . . with a flattish wedge-shaped head like a mini-surfboard.* The actual dimensions are typically 4.5 micrometers long by 2.5 micrometers wide by 1.5 micrometers thick.

177 *The semen pool.* This is actually below the external os if the female
     partner is lying on her back. The cervix relaxes afterwards for the
     external os to dip into the pool.
178 *Seminal plasma gelling.* The gelling reaction occurs when proteins
     from one accessory gland come into contact with enzymes se-
     creted by another accessory gland. In technical terms, the cross-
     linking molecule is ϵ- (γ glutamyl) lysine from the seminal vesicle,
     and the enzyme is a calcium-dependent transglutaminase.
178 *The Billings test and "ferning."* These are described in the *World
     Health Organization Laboratory Manual for the Examination of Human
     Semen and Sperm Cervical Mucus Interaction,* 4th edition, Cambridge
     University Press, Cambridge, 1999.
179 *Ferning.* When I first came across this test, no one could tell me
     what the crystals were. I had them analyzed, and found that they
     were sodium chloride — in other words, common salt, the prin-
     cipal salt in our body fluids. Why, though, should common salt
     form needle-like, branched crystals, rather than the small cubic
     ones that we find in our salt shakers? The answer is that the sur-
     faces of the crystals, identical to the eye, are chemically different.
     Mucopolysaccharides stick preferentially to some faces, prevent-
     ing them from growing further, so the crystal grows unequally in
     different directions. This principle of controlled crystal growth is
     now recognized to permeate nature. It appears to underlie, for ex-
     ample, the development of the shapes of seashells.
        It is unclear why "ferning" relates to the "goodness" of the mu-
     cus. Presumably it is a matter of the hydration and unfolding of
     the mucopolysaccharides, which affects both the consistency of
     the mucus and the ability of the molecules to stick to different
     crystal faces. This is a Ph.D topic waiting for a candidate.
180 *Lateral pressure in a protein film on an oil drop.* Fisher, L. R., Mitchell,
     E. E., and Parker, N. S., "A critical role for interfacial compression
     and coagulation in the stabilisation of emulsions by proteins,"
     Journal of Colloid and Interface Science, vol. 119, 1987, p. 592.
182 *Ability of spermatozoa to fertilize the egg.* It still seems to be an open
     question as to what proportion of the spermatozoa that penetrate
     the cervical mucus are actually capable of fertilizing the egg.
     Robin Baker, in *Sperm Wars* (Basic Books, 1997), puts the propor-
     tion as low as ten percent.
182 *Pushing into cervical mucus.* The force that a sperm swimming at
     3 mm/min can exert is given by:

Force = 6π × radius of head × viscosity of medium × velocity of sperm
    $\approx 10^{-10}$ Newtons

The same result has been obtained by direct experiments, where motile spermatozoa were stuck to a tiny spring and the force generated calculated from the deflection of the spring. The pressure that this force produces is simply the force divided by the cross-sectional area of the head, and comes to 300 Pascals (i.e., about one three-hundredth of atmospheric pressure).

A simplified picture of sperm penetration is that this pressure will enable the swimming spermatozoon to push its way into anything with a yield stress of less than a third of this value, i.e., 100 Pascals.

182 *Effect of female sexual enjoyment on sperm penetration.* Results of a survey by Jacky Boivin of Cardiff University, reported in *New Scientist,* 12 September 1998, p. 20.

182 *Swimming in cervical mucus.* We normally think of swimming in terms of the conservation of momentum. A swimmer "throws" water backwards, and gains an equivalent forward momentum. If the swimmer stops moving the arms, he or she will glide forward under his or her own momentum for several body lengths before being brought to a stop by the viscous drag of the water. Someone carried forward under their own momentum is said to be under the influence of *inertial forces.* The relative importance of inertial forces and viscous forces is given by the *Reynolds number,* which is simply the ratio of the two forces. For a person swimming, the Reynolds number is between ten thousand and a million, which explains why the person can glide forward under their own momentum without being dragged to an abrupt halt by viscous forces. For a swimming spermatozoon, though, the Reynolds number is between 0.1 and 0.01, and viscous forces dominate. Under these conditions, the spermatozoon can only "coast" under its own momentum for a distance given by:

distance = (2 × initial velocity × radius² × density) / viscosity of liquid

This formula tells us that a spermatozoon swimming at 3 mm/min can only "coast" three micrometers in water, i.e., about a twentieth of its own length, before being brought up short by viscous drag on the head. If viscous drag on the now-motionless tail is added in, the distance will be even shorter.

183 *Mucopolysaccharide molecules from bundles.* See review article by Carlstedt, I., and Sheehan, J. K., in *Symposium of the Society for Experimental Biology,* vol. 43, 1989, p. 289.

184 *The shear-thinning behavior of cervical mucus.* This has been reported by Ford, W. C., Ponting, F. A., McLaughlin, E. A., Rees, J. M., and Hull, M. G., in *International Journal of Andrology,* vol. 15, 1992, p. 127.

It is an open guess as to why Nature has made cervical mucus shear-thinning, but one result is that it can stop bacteria penetrating from the vagina to the uterus, since bacteria are too small and their movements are insufficiently vigorous to induce shear-thinning.

184 *Beating of the tails.* See *The Cervix,* Jordan, J. A., Singer, A., Saunders, W. B., and Co. (eds.), London, 1976, p. 169.

184 *Beating of the kinocilia.* Davajan, V., Nakamura, R. M., and Kharma, K., *Obstetric and Gynecological Survey,* vol. 25, 1970, p. 1; Odeblad, E., *Acta Obstetrica et Gynecologica Scandinavia,* vol. 47, 1968, p. 57.

184 *Purcell's article on "Life at Low Reynolds Numbers."* This can be found in *American Journal of Physics,* vol. 45, p. 3.

186 *How human sperm flagellae move.* See article by Phillips, D. M., in *Journal of Cell Biology,* vol. 53, 1972, p. 561.

Bacterial flagellae also use viscous drag to drive themselves forward. The flagellae in this case are rigid helices, attached to a rotating disc driven by a molecular motor at the base of the head (such motors are the only examples that I know of the use of the wheel in Nature).

186 *Volume of the uterus.* The normal volume is incredibly small — about 300 microliters, or one-tenth of a teaspoon. The walls can easily distend, though, to hold a three-kilogram baby.

187 *Surroundings of the egg.* See Cherr, G. N., Yudin, A. L., and Katz, D. F., *Development, Growth and Differentiation,* vol. 32, 1990, p. 353.

coda

Page
190 *Nuclear fission.* The discovery of nuclear fission, reported by Otto Hahn and Fritz Strassmann in the German magazine *Naturwissenschaften* (vol. 27, 1939, p. 11 and p. 89), and the huge amount of energy that could be released was predicted by the theoretical explanation of Lise Meitner and Otto Frisch in *Nature,* vol. 143, 11 February 1939, p. 239.

appendix 1: mayer, joule, and the concept of energy

Page
196 *Joule's experiments.* These were reported by him in *Philosophical Magazine,* series 3, vol. XXVII, 1884, p. 205 (reprinted in Lindsay, R. B., ed., ibid.).

index

Note: Page numbers in *italic* refer to illustrations